A Bite-Sized Public Af

CW00404506

Keeping the Wheels on the Road

UK Auto Post Brexit

Edited by
David Bailey, Alex De Ruyter, Neil Fowler and John Mair

Published by Bite-Sized Books Ltd 2019

Bite-Sized Books Ltd Cleeve Croft, Cleeve Road, Goring RG8 9BJ UK
information@bite-sizedbooks.com
Registered in the UK. Company Registration No: 9395379

Bite-Sized Books Ltd Cleeve Croft, Cleeve Road, Goring RG8 9BJ UK
information@bite-sizedbooks.com
Registered in the UK. Company Registration No: 9395379
ISBN: 9781798005156

Contents

Acknowledgments

This book could not have been more timely; the winter of 2018-19 saw the public meltdown of the British auto industry pre Brexit! It surfs that zeitgeist perfectly.

It is the third in the Bite-Sized Brexit series, steered by Paul Davies, with more to come, and it was possible thanks to the editors, their contacts and their specialist knowledge. The authors, as ever writing for free (as do the editors) make the book what it is.

We are very grateful to all of them and hope this informs the public debate, which has still a long way to go.

David Bailey, Birmingham
Alex De Ruyter, Birmingham (and Australia)
Neil Fowler, Northumberland
John Mair, Oxford

The Editors

David Bailey is Professor of Industrial Strategy at the Aston Business School. He has written extensively on industrial and regional policy, especially in relation to manufacturing and the auto industry. He is now participating in the H2020 RISE project *Makers,* with particular responsibility for the work package on industrial policy. He is a regular newspaper columnist and media commentator. He is the Editor-in-Chief of the journal *Regional Studies*, as well as an editor of *Policy Studies*, and the *Journal of Industry, Competition and Trade.* A previous chair of the Regional Studies Association, David is now chair of the RSA Europe Think-Tank and policy forum. Tweet him @dgbailey

Alex de Ruyter is a professor at Birmingham City University and serves as Director of its Centre for Brexit Studies. He brings a wealth of research experience and academic engagement in the areas of globalisation, regional economic development, labour market and social exclusion issues. He has published more than 60 academic outputs in leading national and international economic journals and has been the recipient of research funding, including being an investigator in an ESRC funded study on the effects on subsequent employment experience of workers from MG Rover after plant closure in 2005.

Professor de Ruyter has undertaken numerous media interviews and is currently researching on the likely impact of Brexit on the UK automotive supply chain in addition to exploring working in the 'gig economy'. He is also a Board member of the Regional Studies Association.

Neil Fowler has been in journalism since graduation, starting life as trainee reporter on the *Leicester Mercury*. He went on to edit four regional dailies, including *The Journal* in the north east of England and *The Western Mail* in Wales. He was then publisher of *The Toronto Sun* in Canada before returning to the UK to edit *Which?* magazine. In 2010/11 he was the Guardian Research Fellow at Oxford University's Nuffield College where he investigated the decline and future of regional and local newspapers in the UK. He remains an Associate Member of Nuffield and now acts as an adviser to organisations on their management and their external and internal communications and media policies and strategies.

John Mair has taught journalism at the Universities of Coventry, Kent, Northampton, Brunel, Edinburgh Napier, Guyana and the Communication University of China. He has edited 28 'hackademic' volumes over the last eight years on subjects ranging from trust in television, the health of investigative journalism, reporting the 'Arab Spring', to three volumes on the Leveson Inquiry. John also created the Coventry Conversations, which attracted 350 media movers and shakers to Coventry University. Since then, he has launched the Northampton Chronicles, Media Mondays at Napier and most recently the Harrow Conversations at Westminster University. In a previous life, he was an award-winning producer/director for the BBC, ITV and Channel 4, and a secondary school teacher.

Introduction

On the road but going where?

Neil Fowler

The UK car industry is in crisis. There can be no doubt about that. But the exact causes are a matter of intense debate.

Is it a perfect storm of diesel phobia, electric adulation, urban congestion and millennial minimalism that has led to a sudden downturn in our near century-long love affair with the motor car?

And is Brexit the final factor that has come along to confirm that all is desperate in the automotive world? Or is Brexit, and the uncertainty that has prevailed over the past 30 months or so around it, the principal cause of this current malaise?

This book seeks to answer the underlying elements that have created this maelstrom – and the contributors offer real and often opposing insights in to what has been happening and where the sector may end up.

Justin Cox and David Oakley of LMC Automotive start us off by forensically analysing the statistics – which don't make for pretty reading. They are pessimistic on what may happen: "It would be overly optimistic to assume that manufacturers which export large portions of their UK output to the EU would continue to do so indefinitely under persistently unprofitable operating conditions," they say. "Indeed, should more of these currently UK-produced models fail to be renewed in Britain, annual UK car production could be over 500,000 units lower in the second half of the next decade when compared with our 'base' view."

Ross Clark is much more optimistic, though he is scathing on the failings of the big companies:

"When Nissan announced it would not, after all, produce its new X-Trail in Sunderland, this was reported as proof of an impending Brexit disaster. A Labour councillor in South Wales even suggested that 'all those who voted to leave should be laid off first'. But Nissan's decision has little to do with Brexit, and everything to do with the turmoil of the global car industry, " he writes.

"It is not that overall car sales are plunging — they grew by a modest 0.5 per cent across Europe in 2018. The problem is that established carmakers have failed to keep up, and their future now looks far more uncertain than it did even just a few years ago."

Ian Henry of Auto Analysis is another who has delved in to the figures and who has been listening to what some outsiders have been saying. "Investment in the UK has fallen substantially in the last couple of years. Further decline will follow, maybe rapidly," he predicts.

"The calm and measured words of Japanese diplomats in the UK are worth taking on board: leaving the EU means that the UK's role as 'gateway to Europe' is over. A hard Brexit means the gate will be closed more quickly than otherwise would be the case." He forecasts a fall of output of at least 175,000 cars a year in the event of no deal. There would be worse to come longer term.

Leading economist Vicky Pryce has a detailed examination of what might happen to government help in the post-EU world.

"We know that in general the UK has tended to receive more back in science money than it has contributed through the EU budget process," she writes. "If that source of future funding is cut or curtailed, will the Government, already having to borrow much more than it had envisaged before the Referendum, be able to step in?" she asks. "And will the Treasury have enough resources or the willingness to assist the regions that are likely to suffer most, both from the reduction in EU transfers and skilled labour and the damage that may be done to particular sectors such as the automotive industry by a badly executed Brexit? Don't hold your breath."

David Bailey, co-editor of this book, and Richard Burden MP look at the potential effects of a hard Brexit, combined with confused automotive policies, might have on the substantial Midlands car industry. They mix hope and fear.

"Make no mistake, the innovative capacity and diversity that has made the UK automotive sector the success story it has become over the past decade remain in place, they write. "However, the fact that a number of major manufacturers have yet to confirm plans to build in the UK the next generations of models (that the UK should be in pole position to produce) sends out serious warnings signals that would be foolish in the extreme to ignore."

David Shepherd also knows the Midlands well and has witnessed the region's assorted auto traumas in his career. But he is more worried this time: "Over the years, the automotive industry has weathered many storms but this time it feels different — and that's because it is different. This time the storm is unique to the UK and, somehow, it feels like we are on an uneven playing field. This isn't part of a cycle we've seen in the industry before and needs to be treated differently."

Neil Winton, also a long-term observer, takes a more positive view, though, echoing much of what Ross Clark says. "Britain's automotive business will thrive after the EU exit," he says. "The Referendum vote is being used to mask the sector's poor strategy and vision. The canard that ports will jam and motorways come to a halt the day after Brexit is just that, from an orchestrated industry response that is fearful of any change to its cosy arrangements."

David Bailey, in his second chapter, asks for an upgrading in how the UK develops its future manufacturing plans. "There is a strong case for UK industrial strategy to be afforded an institutional status similar to both UK monetary and fiscal policies," he says. "At the very least, it should be the subject of regular strategic long-term reviews. By giving it that sort of priority, the new government would send out the kind of powerful message that British industry and foreign investors need to hear given recent uncertainty."

None of this is a panacea for a messy Brexit, he writes. "But regardless of the form of Brexit, a more interventionist industrial policy will be required for UK auto going forward, building on the public-private cooperation that has been developed so well over the last decade, and in contrast to what some deregulation-minded Brexiteers might suggest."

Unite is the principal trade union in the automotive industry — and in a hard-hitting chapter Steve Turner and Ben Norman say that Unite will defend its members and the industry from fundamental change.

"The talent, creativity and ingenuity of automotive workers in the UK is beyond question," they write. "Nor is their determination. Hard work, flexibility and even sacrifice rebuilt the industry. This will not be allowed to be thrown away by fantasists and ideologues. Unite will continue to campaign for the future of the automotive industry."

The book concludes with a sobering report from Australia, written by co-editor Alex De Ruyter. In 1970 the country was producing some 475,000 cars a year, he writes. But from then on, for various reasons, decline set in. "By the end of 2017 a period of some 70 years of large-scale vehicle production in Australia had come to an end, he says."

"For the Brexiting Britain, the parallels are striking. Like Australia in its protectionist days, domestic production was feasible when the presence of external trade barriers acted as an inhibitor to exporting to the UK from a country of origin outside the EU; and the location-specific advantages of the UK with its flexible market environment *inside* the EU made it an attractive location to be a production platform integrated with the rest of the EU.

"Stripped of these advantages post-Brexit and facing likely new tariff and regulatory barriers, the clear incentive – as Honda and Nissan are currently demonstrating – for multinationals, will be to divest themselves over the coming period of production in the UK and reinvest elsewhere."

Chapter 1

Overt optimism cannot hide the facts

The macroeconomic impact on the auto sector could be devastating, say Justin Cox and David Oakley of LMC Automotive

As we write, the terms of the UK's withdrawal from the EU are still highly uncertain. The possibilities range from a disorderly, no-deal Brexit, in which the UK reverts to trading on World Trade Organisation (WTO) rules, to something closer to the deal agreed by Theresa May in November 2018, but subsequently rejected by Parliament – which would preserve existing trading relationships for a transition period until December 2020, after which it is hoped a comprehensive free-trade agreement will be in place.

The latter proposal has thus far been met with overwhelming hostility from MPs, with the Government losing the meaningful vote on the deal in the House of Commons by a record 230-vote margin. The most contentious issue was – and remains – the so-called backstop arrangement to avoid a hard border on the island of Ireland.

This is in large part due to the fact that Northern Ireland would effectively remain in the EU's Customs Union after December 2020 if a free-trade agreement were not signed by that time, and only the EU would have the right to decide when this state of affairs could be ended. Although Prime Minister May has pledged to secure changes to the backstop, the EU remains unwilling to reopen the legally-binding withdrawal agreement, so it is still unclear whether sufficient concessions will be made on either side in order for progress to be achieved on this issue.

On the other hand, a variety of alternative scenarios cannot be ruled out at this stage. These would include Article 50 being extended to allow more time for negotiations – or possibly preparations for a 'managed no-deal' Brexit – to take place; a second referendum being held, possibly resulting in Brexit being cancelled altogether; or a General Election being called.

However, in the analysis below, we focus on what currently appear to be the most likely options of either a modified version of the deal agreed with the EU eventually being approved by Parliament, or the UK leaving the EU without a deal.

A deal is likely

Despite seemingly discouraging recent developments, our partners at Oxford Economics (OE) still believe that the most probable scenario is that MPs will ultimately vote in favour of a deal, and therefore this informs both OE's and LMC's baseline scenarios. From a macroeconomic point of view, we assume that there would be little difference in the short term between the deal currently being proposed and a scenario in which the UK remained in the EU.

In the base case scenario, OE sees UK GDP growth at 1.4 per cent in 2019, and then an average of 2 per cent in 2020-22. The relatively buoyant medium-term forecast reflects a boost from pent-up demand due to the uncertainty created by Brexit, as well as some fiscal loosening from the Treasury as announced in the 2018 Budget.

Under a no-deal scenario, OE is not forecasting that the UK enters recession, although two consecutive quarters of negative growth cannot be ruled out. For 2019 as a whole, OE forecasts GDP growth of 1 per cent, followed by 0.8 per cent in 2020, before a pick-up to 1.5 per cent in 2021. The worst potential effects of no deal would be mitigated, OE assumes, by higher government spending – and therefore a greater toleration for a higher deficit, at least temporarily – as well as a cut in interest rates, to 0.25 per cent. This interest rate cut would be delivered in spite of a spike in CPI inflation to around 4 per cent in the second half of 2019.

The impact on sales

UK light vehicle (LV) sales – which include both personal vehicles and light commercial vehicles – totalled 2.73m units in 2018. Whilst this represents a 6.1 per cent fall year-on-year (YoY), it should be noted that the UK market is receding from an all-time record of 3.08m units in 2016, and a cyclical slowdown would be expected in any event. In addition, Brexit is certainly not the only factor that has been affecting sales in recent times, although uncertainty around the future of the economy was increasingly cited as a cause of slowing sales as 2018 progressed.

The introduction of a new emissions testing procedure, known as WLTP, disrupted sales across Europe in 2018, with purchases pulled forward to August before falling away sharply in September and subsequent months, with sales still struggling to fully recover by year-end. Furthermore, diesel sales have declined severely amidst a spate of negative publicity for the fuel type, linked to harmful emissions and therefore potential bans in the future. Early indications for 2019 suggest that the UK market is recovering from the impact of WLTP, but sales are still subdued, in part due to low levels of consumer and business confidence.

Under our baseline scenario, which assumes an orderly exit from the EU, LV sales are set to grow in the coming years, although that growth will be modest, with 2019 LV sales reaching 2.76m, a 1.2 per cent increase YoY, followed by a 1.5 per cent gain in 2020 and 2 per cent in 2021, by which time LV sales will be 2.86m.

However, under a no-deal scenario, a number of macroeconomic factors would conspire to put downward pressure on sales. As well as a reduction in GDP growth mentioned above, the pound would depreciate sharply, to – OE expects – $1.13 and €1.02 in Q2 2019. This would increase the cost of vehicles in the UK, given that most cars sold in the UK are imported (predominantly from the EU), and even those that are locally manufactured rely on international supply chains.

Add to this, a 10 per cent WTO tariff to be applied on the imported vehicles, and the pressure on prices begins to really accumulate. Supply chains would also be affected by 10 per cent tariffs in the event of a no deal, as well as non-tariff barriers such as increased customs checks at the border, all of which will increase cost pressures on manufacturers. After sterling's depreciation in the wake of the 2016 referendum, automakers absorbed some of the costs stemming from the weaker pound, but would surely struggle to do so any further were the currency to slip to the levels predicted by OE, and tariffs were applied. It has recently been reported that Porsche is asking UK buyers to agree to pay an additional charge of up to 10 per cent of the purchase price of a new vehicle in the event of a no-deal Brexit.

We expect that the combined impact of these effects would be to reduce UK LV sales by around 10 per cent by 2020, compared to the baseline scenario. That is, sales would fall 5.3 per cent YoY in 2019, to 2.59m, followed by a further 2.1 per cent YoY decline in 2020, to 2.53m. This compares to the 2.8m sales in 2020 envisaged in the base case. Over the 2019 to 2024 period, we forecast that a cumulative 1.4m UK sales could be lost in a no-deal scenario, versus the base case.

Although the effect on other countries would be much less significant, we still see global sales being reduced by some 200,000 units per year in the short-term, in addition to those lost in the UK, if no deal is agreed.

The impact on production

The auto industries in the UK and Europe are closely integrated. Common regulatory EU-wide frameworks have evolved to facilitate the development of highly complex automotive supply chains, which now underpin a business model that relies on just-in-time and just-in-sequence delivery and production.

This business model will be affected by Brexit, but the scale of impact depends on which scenario is pursued: our 'base' transitional Brexit or the 'no-deal' scenario.

UK LV production is expected to have reached 1.6m units in 2018 and, of this total, 80 per cent is exported globally, with the EU receiving more than 55 per cent of this export volume. However, it is important to appreciate the complexity of automotive manufacturing with each finished vehicle consisting of thousands of components, which in turn are processed and traded across multiple borders. For UK production, 44 per cent of a vehicle's content is sourced locally, with almost 80 per cent of the remaining non-domestic content coming from the EU.

In a no-deal Brexit, barrier-free access to the EU market is expected to end with the UK withdrawing from the EU on WTO terms. Under these terms, a 10 per cent tariff would be applied to all traded goods. Coupled with regulatory requirements, these new customs procedures will inevitably add barriers to trade, increase red-tape and ultimately cost. Border checks in particular will cause delays and undermine the efficiency of the just-in-time model.

Although the shape of Brexit remains unclear, UK-based OEMs are already putting in place no-deal contingency plans to increase stocks of components, as well as finished vehicles, so that any supply-chain disruptions can be minimised. While this may temporarily provide a boost to vehicle output in the opening quarter, we expect this to be unwound in the event of a 'hard' Brexit due to weaker UK market demand and fading competitiveness. UK LV production could fall to 1.4m in 2019 and 1.48m in 2020 – respectively 11 per cent and 7 per cent lower than our 'base' transitional Brexit.

Longer term, continuing obstacles to trade associated with a no-deal Brexit may encourage an increase in UK component manufacturing 'onshoring'. However, fears persist that UK domestic OEMs lack the required local market scale to attract investment by key volume component suppliers – very simply, large capacity investment decisions are more likely to gravitate to the EU where the combined potential of the EU Single Market offers the greatest volume opportunity and manufacturing economies. This is particularly relevant for an industry which is facing such technological change as all automakers grapple with huge new investment decisions required to accommodate the evolution of electrification and autonomous driving.

For UK-based OEMs, a no-deal Brexit might even threaten the existence of certain assembly facilities. Indeed, the fear of a no-deal scenario may have contributed to the recent announcements by Nissan to cancel the expansion plans of its Sunderland plant, and Honda, which has confirmed that it will close its UK manufacturing operation in Swindon from 2021.

Admittedly, many of these decisions are based upon the new EU-Japan trading accord. In force since 1 February 2019, this new trade agreement ensures that the EU's 10 per cent tariff on Japanese auto imports will be tapered to zero over the next ten years.

Increasingly Japanese OEMs may choose to import new models rather than invest and continue to localise production in their European 'transplants'. For the UK, whose access to the EU's Single Market originally attracted Japanese automakers, Brexit adds another dimension. Should the UK leave without a deal and WTO tariffs are applied to UK vehicle exports, the same cars made in Japan may well end up costing less to import into the EU27 than those produced just across the Channel in Britain.

With the UK's Japanese transplants collectively producing almost half of the UK's Light Vehicle output in 2018, and Toyota and Nissan exporting 53 per cent and 90 per cent of their production to the EU, risks for further industrial shrinkage are significant. For PSA Group, which is battling with overcapacity, the Opel/Vauxhall Ellesmere Port plant may be tough to justify after 2021, when the current-generation Astra compact model produced there is slated for replacement. Jaguar Land Rover, meanwhile, could divert more production to a new facility in Slovakia and/or increase contract production with Magna in Austria.

As a result, it would be overly optimistic to assume that manufacturers which export large portions of their UK output to the EU would continue to do so indefinitely under persistently unprofitable operating conditions. Indeed, should more of these currently UK-produced models fail to be renewed in Britain, annual UK car production could be over 500,000 units lower in the second half of the next decade when compared with our 'base' view.

About the contributors

Justin Cox is Director, Global Production, at LMC Automotive. In this role he provides the functional lead and global oversight of the company's automotive production forecast activity. With more than 20 years of extensive automotive business experience, including OEM and supplier roles, he has gained a 'real world' insight into the commercial pressures and planning challenges facing the industry throughout the value-chain. Justin holds a BSc degree in Economics and Human Geography from the University of Reading.

David Oakley is EMEA Analyst at LMC Automotive. In this role he oversees Light Vehicle sales forecasting for the Middle East and Africa, as well as selected markets across Western and Eastern Europe. He holds a First-Class Honours degree in Politics from the University of Nottingham, and enjoys applying political and economic analysis to the automotive industry.

LMC Automotive is the leading independent and exclusively automotive focused provider of global forecasting and market intelligence in the areas of vehicle sales, production, powertrains and electrification. Highly respected for its responsive customer support, the company's client base from around the globe includes car and truck makers, component manufacturers and suppliers, financial, logistics and government institutions.

Chapter 2

Carmageddon: the auto industry struggling in a changing world

The motor giants are facing a challenging future which has little to do with Brexit, says Ross Clark

When Nissan announced it would not, after all, produce its new X-Trail in Sunderland, this was reported as proof of an impending Brexit disaster. A Labour councillor in South Wales even suggested that 'all those who voted to leave should be laid off first'. But Nissan's decision has little to do with Brexit, and everything to do with the turmoil of the global car industry.

It is not that overall car sales are plunging — they grew by a modest 0.5 per cent across Europe in 2018. The problem is that established carmakers have failed to keep up, and their future now looks far more uncertain than it did even just a few years ago.

BMW, Mercedes, Volkswagen, Nissan: for decades, the same names ruled. It was a complacent industry, and progress was incremental. Every five years or so, a new model of car would be brought out that was slightly better, slightly more efficient than the last. The domination of the internal combustion engine meant that this piece of late 19th-century technology set a huge entry barrier to new entrants. You couldn't set up a car company from scratch and hope to steal a march on the established players.

So they scoffed at suggestions that their world might be upended by electronic cars, ride-sharing apps like Uber — which could mean fewer people owning cars — or various degrees of driverless technology.

Upstarts upset the market

But life has come at them hard. Tesla has proved that there is a large market (and long waiting lists) for premium models — and, so far, the company has defied the short-sellers who have bet on its demise. As for self-driving technology, it is Google which has led the way, investing more than a billion dollars. A fatal accident last year hasn't deterred Uber either. Eager to catch up, Toyota recently invested $500m in Uber to help develop self-driving technology.

Another upstart in the automotive sector is Dyson, which since 2015 has been investing £2.5bn to develop, in Wiltshire, an electric car with some element of driverless control. Driverless technology is crucial for the future shape of the automotive sector because it promises to slash the cost of taking a taxi — which could undermine car ownership altogether, at least in cities.

Alternatively, it may turn out that hydrogen fuel cells prove to be the low-emission, carbon-neutral long-term replacement for petrol and diesel. Toyota is certainly looking that way — it recently launched a hydrogen car, the Mirai. Hyundai, too, is investing heavily in hydrogen, last December announcing a £5.5bn investment in the technology The battle between the two — hydrogen vs electric — has been dubbed by analysts at KPMG the car industry's 'Betamax vs VHS' moment, echoing the big battle over video technology in the 1980s.

Electric cars ought to be more efficient, as hydrogen first has to be extracted from water — there being no natural earthly source for pure hydrogen. But then, unlike battery cars, hydrogen cells can be refuelled in minutes. That is one reason why, according to a survey by KPMG, motor executives believe battery cars will eventually lose the battle. But nobody knows, which is what makes the future so uncertain for the car industry. And we all know how the video battle played out — a victory for VHS before it, in turn, was blown away by DVD.

Governments killing off the diesel

But the bigger factor in the car giants' malaise is the way governments have sought to speed up the car revolution by killing off traditional engines. What did for the Sunderland X-Trail — a big, bad diesel car — was not Brexit but Michael Gove's announcement in July 2017, echoed in France, that all new diesel and petrol cars will be banned from 2040. Then came initiatives like London's Ultra-Low Emissions zone, which will impose new charges on older diesel vehicles entering the congestion charge zone from this year.

You can see versions of this all over the western world. A recent German court ruling allowed cities to ban diesel cars completely. With restrictions in Belgium and the Netherlands too, motorists are becoming reluctant to buy them. On top of this, Nissan had been struggling to make its diesel X-Trail compliant with the latest EU emissions regulations. When the company made its pledge in 2016, things looked very different than they do now.

The technology now on the way — be it battery, hydrogen cell or something else — is not necessarily going to be developed by the old players. Meanwhile, the old companies are being shaken by scandal. Martin Winterkorn had to quit as chairman of Volkswagen after the company was found to have cheated in emissions tests when it was struggling to develop cleaner engines. In late 2018, BMW was fined $10m over its handling of a spate of engine fires in South Korea. Carlos Ghosn, the Nissan chief executive who made the decision to build the X-Trail in Sunderland, is currently on charges for fraud, and the company is to be run by a more Japan-centric management.

Cities are also becoming hostile environments for cars — with congestion charges and ever-tighter parking restrictions. It's not that cars will disappear — but in future, urbanites will be far less likely to own them and more likely to hire them on an intermittent basis. While the number of UK residents with a driving licence continues to creep upwards, there has been a huge shift in the age profile over the past quarter century. In 1994, half of 17- to 20-year-olds could drive; now it's less than a third. How many of the Uber generation will ever own a car? No one knows, but if you are a car manufacturer whose business model assumes the continued growth of what Mrs Thatcher liked to call 'the great car economy', with two cars on the driveway of the archetypal middle-class home, it is enough to put your strategy in a spin.

Factories closing around the world

As manufacturers wrestle with the consequences of these various trends, factories are closing at an increasing rate. GM is closing five plants across the US and one in Canada as it seeks to invest more in electric vehicles. Ford has just cut 1,600 jobs in Germany as it, too, seeks to invest more in electric cars. The people of Sunderland — where, by the way, Nissan is not proposing to cut a single job, only to cancel a planned expansion of its existing plant — are the lucky ones. Nissan's factory there is one of the most efficient in the world and not so long ago was producing more cars than the whole of Italy.

No one yet knows who the winners of Carmageddon will be, but for the moment initiative seems to be slipping away eastwards. China, until recently damned as the dirty man of the world for its fondness for opening new coal-fired power stations, has taken to electric vehicles with an enthusiasm greater than anywhere else. In 2018, the Chinese bought a million electric cars, a 60 per cent increase in two years — this in spite of an overall fall in car sales in the country. That is why James Dyson has relocated his company HQ to Singapore — to be closer to the epicentre of the electric car revolution.

Western luxury carmakers, who for years have relied on selling their products to a fast-expanding Chinese middle class, are suddenly looking flat-footed. It was principally falling Chinese demand which led Jaguar Land Rover to announce in January that it is chopping 5,000, or one in eight, UK jobs — although inevitably that too was mistakenly interpreted as a consequence of Brexit.

It is easy to exaggerate trends. While diesel sales are in steep decline, electric vehicles still account for no more than 2 per cent of new vehicle registrations in most European countries (Norway, on nearly 40 per cent, is an outlier). By some estimates, at the end of the next decade, one in five cars bought in America will be electric. But for the moment, it is petrol and petrol hybrids which are filling the gap as consumers turn away from diesel (sales of which were down 18 per cent in the third quarter of last year, while petrol sales rose 15 per cent).

Battery technology is developing fast, and the price of the batteries falling fast, but it may never advance to the stage where electric cars are a viable alternative to petrol and diesel for long journeys in remote areas. And Emmanuel Macron is learning the hard way what happens to politicians who get too carried away with green taxes: the rise of the Gilets Jaunes has shown the hazards of unleashing policies designed around the needs of cities on to rural populations, for whom, for the moment, old-fashioned cars provide the sole means of practical transport.

A head start for Sunderland

But if the electric car is to be part of the future, Nissan and Sunderland have a head start, because for several years they have been making the world's number one electric car for the mass market, the Leaf — a vehicle whose development was championed by the fallen Carlos Ghosn. The environmentally conscious would surely prefer this vehicle to be made in

Britain in place of the X-Trail. Other carmakers are now rushing to catch up: Volkswagen is spending $40bn over the next five years to electrify its 300-odd models.

There is no guarantee that either Nissan or its Sunderland plant will be around in 20 years. Today's electric vehicles may by then be museum pieces, driven out by hydrogen or something we have yet to discover. But whatever technology does come to take over the car market, you can be sure that it will have been developed in an economy that supports innovation, and which has the lowest possible barriers to hiring the required talent from around the world.

So in that sense, the Nissan story is related to Brexit. It's a reminder that the world economy is fast changing — and that success will go to whoever can step up a gear.

About the contributor

Ross Clark is a freelance journalist. This chapter first appeared as an article in *The Spectator*.

Chapter 3

What cost a hard Brexit?

Leaving the EU with no deal would increase the automotive industry's costs significantly, by more than £3bn a year through tariffs alone; further costs would arise through non-tariff barriers and other general disruption. Ian Henry of AutoAnalysis believes strategies to avoid tariffs will have only limited impact, with a hard Brexit also leading to reduced UK vehicle production by at least 175,000 a year, declining exports to the EU and falling investment

As the UK's departure from the EU nears, future trading relationships remain uncertain. There is a proposed but as yet unratified withdrawal agreement, during which nothing substantive would change until the end of 2020. The UK automotive industry does not want to leave[1] but sees the transition period as an opportunity to prepare for life outside the EU.

A hard Brexit would mean the UK and EU switching immediately to trading on WTO terms. Vehicles exported to the EU would face a 10 per cent tariff, with the same rate applied to EU exports to the UK. Tariffs would also apply to other markets with which the EU has trade deals: the UK would lose the benefits of such deals once outside the EU. In addition, components would be subject to tariffs within a range of 2.5-4.5 per cent; some components or materials, notably aluminium sheet used by Jaguar Land Rover (JLR), would be subject to higher rates.

As well as tariffs, a hard Brexit would see border delays as customs and other regulatory checks are introduced. While these may be partly obviated through schemes such as Authorised Economic Operator status and advanced submission of relevant documentation, the industry's transporters would be caught up in the flow of other sectors without such status. The sector's finely-tuned just-in-time supply chains will be at risk.[2]

UK automotive manufacturing has experience of tariffs on components sourced outside the EU, or on vehicle exports beyond the EU. For example, vehicles shipped to the US are subject to import duty of 2.5 per cent.[3] Until the EU-Japan Free Trade Agreement came into effect at the start of February 2019, cars and components traded between the EU and Japan remained subject to the 10 per cent tariff on cars and the lower rate on components.[4] However, adding tariffs to EU trade would be on a far greater scale and place strain on its long-term economic viability from which it may not recover.

Calculating the cost of no deal

AutoAnalysis has developed a model to calculate the impact of tariffs on UK vehicle manufacturing. This produces results broadly consistent with public statements by the industry. JLR has said that it expects a hard Brexit tariff cost of £1.2bn.[5] The model uses public data, alongside proprietary information which AutoAnalysis has collected in recent years. The model works as follows, taking a hypothetical UK-made vehicle:[6] Annual production: 200,000, with 150,000 EU exports:

- Finished vehicle EU landed cost subject to duty: £21,000
- External bill of materials: £9,000, of which EU sourcing is 50 per cent at tier 1s; UK content is 40 per cent at tier 1s, but 30 per cent of this is actually sourced from the EU
- Average imported components tariff: 3.5 per cent
- EU finished vehicle tariff: 10 per cent

Taking these base parameters, the tables below show how tariffs would add slightly more than £39m to the bill of materials cost, but generate a far higher number, £315m, in tariffs on finished vehicles.[7] The total can be reduced through forward processing relief: tariffs paid on component imports can be offset against tariffs paid on vehicle exports. Here, with 75 per cent of production exported to the EU, 75 per cent of the tariffs paid on imported components (£29.3m) could be reclaimed. However, setting up HMRC-compliant systems to achieve this is neither simple nor quick.[8]

Imported component tariff cost for a hypothetical UK-made vehicle

Import costs	£
Bill of materials	9000
EU sourcing 50%	4500
UK sourcing 40%	3600
Of which 30% from EU	1080
Total EU content per vehicle	5580
Tariff cost per vehicle (3.5%)	195.3
Total cost (200,000 vehicles)	39,060,000

Export tariff cost for a hypothetical UK-made vehicle

Export costs	£
EU landed cost of vehicle, subject to duty	21000
Tariff payable per vehicle (10%)	2100
Total cost (150,000 vehicles exported to EU)	315,000,000
"New" cost of vehicle pre dealer margin etc	23100

We have applied the above process to each volume vehicle manufacturer in the UK using the average of their production volumes and EU exports for 2017 and 2018. While we cannot reveal individual company figures, we can say that the costs of tariffs range from c£125m to more than £1bn per company. The total annual cost to the UK's volume vehicle companies is at least £2.7bn. Similar calculations for the lower volume manufacturers, Ford's UK engine manufacturing operations and the engine operations of the Japanese vehicle companies, which export a proportion of their UK-made engines, generate additional tariff costs of more than £400m a year.

In practice, a hard Brexit would add more than £3bn to UK vehicle manufacturing's annual costs through tariffs alone. This is unsustainable. For example, in 2016/2017, JLR made profits of £1.6bn, and £1.5bn in 2017/2018. JLR will break even at best in 2018/19 (it has already reported losses in three quarters) although the reasons for this are varied, and are not solely due to Brexit.[9] It is clear that tariffs would eradicate most, if not all, of the company's profits. Ford has referred to a hard Brexit as a 'red line' of its own, an ominous warning for the long-term future of the company's UK operations.[10]

In addition to identifiable tariff costs, the industry would face non-tariff barriers (NTBs), covering the costs of regulatory compliance, customs processing, additional delays at the borders, the need for additional stocks to cover supply chain disruption and increased working capital requirements to cover additional stocks, funding tariff payments and changes to the VAT regime. Research from other parts of the world suggest that NTBs in automotive could amount to up to 10 per cent.[11]

Countering these costs

There are no simple solutions. If the UK Government wanted to refund tariffs to affected companies and industries, this would place an unsupportable burden on the public purse. More significantly, such a strategy would not be allowed under WTO rules. There has been some talk of a temporary zero-tariff arrangement; pro-Brexit politicians have suggested that Article 24 of GATT could be invoked permitting zero tariffs ahead of a long-term FTA.[12] There is no certainty that this would be accepted or practical (the EU would have to agree for one thing) and we doubt this would be achievable in practice.[13]

However, there are some generic mitigation options for the industry including:

- **Increasing UK content of UK-made vehicles**: this is fine in principle but financial and management resource limitations of the supply base mean this would likely not happen on a sufficient scale or speed to eliminate the cost of tariffs. Complex automotive supply chains mean that simply relocating tier 1 suppliers to the UK would not suffice. Companies throughout the supply chain at tier 2, 3 and 4 levels would also need to relocate to the UK to meet FTA rule of origin requirements.[14] Automotive suppliers want to be near their largest addressable market, and this generally means locating in the EU, and not the UK.[15]
- **Relocating UK production to factories inside the EU**: such moves would damage the UK economy, and while we think this will likely happen over time, vehicle companies do not all have spare EU capacity readily available to permit this taking place quickly or without significant cost.
- **Changing the markets into which UK manufacturers sell**: this too is easier said than done. JLR and Mini already export worldwide and it is questionable whether lost EU sales (caused by tariff-induced price rises) could be recovered through additional non-EU sales. Toyota and Vauxhall typically ship 80-85 per cent of their output to the EU and other markets in wider Europe (Turkey, Switzerland, Norway etc); it is

difficult to see which markets UK factories could be allocated without affecting operations elsewhere.

- **Refocusing UK production to supply the domestic market**: this may be possible but would require investment to enable UK factories to make a wider range of models[16]; most of the best-selling models in the UK are not made here.[17]

Car companies' Brexit amelioration plans

The car companies have begun to implement contingency or amelioration plans to counter a no-deal Brexit:

BMW will close Mini for all of April; it was planning to close the factory in August to prepare for the new electric Mini and update the paint shop but will bring this temporary closure forward.[18]

Honda has announced six non-production days in April to avoid potential supply chain disruption.[19] JLR will extend its annual April shutdown, for the same reason.[20]

Others are increasing UK-held stocks. The warehousing implications of this are significant and with other industries such as food also taking on additional space, there is a limit on how much can be achieved in this way.[21]

Aston Martin will import through ports other than Dover and will also, where necessary, fly components in, though this is clearly an expensive operation.[22]

Conclusion

A hard Brexit would increase the industry's costs by more than £3bn a year through tariffs alone. There are some amelioration strategies which the industry can adopt but these are unlikely to solve the long-run problems that could arise. We expect around 10 per cent of 2018 output would be lost in 2019 even with a soft Brexit due to known production shutdowns and planned model changes; lost production this year would reach **at least** 175,000 units in the event of no deal, and much more in the long run.[23] A hard Brexit would result in a smaller and much less dynamic industry in the UK, with reduced sales into the EU and reduced investment in the future.

Investment in the UK has fallen substantially in the last couple of years.[24] Further decline will follow, maybe rapidly.[25] The calm and measured words of Japanese diplomats in the UK are worth taking on board: leaving the EU means that the UK's role as 'gateway to Europe' is over.[26] A hard Brexit means the gate will be closed more quickly than otherwise would be the case.

Notes

1. https://www.smmt.co.uk/2018/11/no-deal-brexit-catastrophic-blow-to-british-auto-industry-warn-businesses-in-new-survey/

2. For example, see Toyota's views at: https://www.ft.com/content/7260069a-d392-11e8-a9f2-7574db66bcd5

3. Press commentators typically report that US vehicle import duty is 10 per cent; this is incorrect: see: http://trade.ec.europa.eu/doclib/docs/2015/january/tradoc_152998.1 per cent20Trade per cent20in per cent20goods per cent20and per cent20customs per cent20tariffs.pdf

4. http://ec.europa.eu/trade/policy/in-focus/eu-japan-economic-partnership-agreement/

5. https://www.just-auto.com/news/paris-bad-brexit-could-cost-jlr-gbp12bn_id184835.aspx

6. An early version of this was described in:; https://www.just-auto.com/analysis/brexits-come-home-how-the-costs-mount-up_id183422.aspx

7. It is important to note that the import duty is not applied to the retail price of the vehicle but to the lower landed cost of the vehicle at the port of entry.

8. One car company which has set up this system told us in a personal communication that it took "more than a couple of years" and cost "single figure millions" to set up HMRC-compliant systems to recoup duties paid on imported components

9. https://media.jaguarlandrover.com/news/2017/05/jaguar-land-rover-finishes-year-record-sales-volumes-and-revenue-after-strong-fourth, https://media.jaguarlandrover.com/news/2018/05/jaguar-land-rover-reports-full-year-results and https://www.jaguarlandrover.com/news/2018/10/jaguar-land-rover-reports-second-quarter-results-20182019-financial-year. JLR's sales have fallen owing to combination of Chinese market downturn, concerns over diesel in Europe, disappointing sales of Jaguar saloons and the switch to a new Evoque, with production of the old model ending during 2018. Its Q3 figures include a substantial write down of assets as well as further deterioration in trading performance: https://media.jaguarlandrover.com/news/2019/02/jaguar-land-rover-reports-third-quarter-results-20182019-financial-year

10. https://www.just-auto.com/news/ford-warns-that-hard-brexit-is-a-red-line_id185097.aspx -

11. Cadot and Gourdon (2016), p 241

12. For example see: https://brexitcentral.com/managed-no-deal-wto-option-using-article-24-gatt-can-avoid-raising-tariffs-quotas/

13. https://uk.reuters.com/article/us-britain-eu-wto/trade-experts-dismiss-brexiteers-faith-in-obscure-wto-clause-idUKKCN1PH24V; https://uktradeforum.net/2019/01/26/why-claims-about-a-wto-article-24-interim-agreement-are-a-red-herring/;

https://tradebetablog.wordpress.com/2018/12/27/myth-of-10-year-grace-period/ all explain why the article 24 route will almost certainly not work, and indeed is inappropriate

14. FTAs typically require local originating content of 55-60% and no UK-made car meets this criterion

15. During Q1.2019, AutoAnalysis has been told by a number of UK suppliers that their EU-based vehicle company customers want them to transfer production tooling to the EU to avoid supply disruption from the UK to the EU. This only further increases the pressure on the industry.

16. BMW is reported to be considering adjusting its Oxford plant to make some of the 1- and 2-series models for the UK, alongside Minis: see Autocar magazine, February 6, 2019

17. The data for 2018 shows that only three of the top ten sellers in the UK (Mini, Astra and Qashqai) were made in the UK, and some of the Minis were actually not made in the UK. https://www.smmt.co.uk/vehicle-data/car-registrations/

18. https://www.bbc.co.uk/news/business-45561908

19. https://europe.autonews.com/automakers/honda-plans-front-load-uk-production-ahead-brexit

20. https://www.bbc.co.uk/news/business-46993480

21. Significant detail on Honda's views on how a Hard Brexit would impact its UK operations available at: https://www.ft.com/content/8f46b0d4-77b6-11e8-8e67-1e1a0846c475

22. https://www.theguardian.com/business/2018/nov/15/aston-martin-fly-car-parts-uk-brexit-deal-collapses-sales

23. AutoAnalysis produces the SMMT's Production Outlook report and has set out its scenarios for reduced production in its recent reports which are available from SMMT at: https://www.smmt.co.uk/vehicle-data/production-outlook/

24. https://www.smmt.co.uk/2019/01/uk-automotive-on-red-alert-as-no-deal-threat-sees-manufacturing-and-investment-plummet/ reports that investment in UK automotive fell by 46.5 per cent in 2018 alone

25. https://www.independent.co.uk/voices/brexit-latest-nissan-xtrail-sunderland-car-industry-bmw-mini-toyota-honda-peugeot-vauxhall-a8760361.html. This view was backed up by an AutoAnalysis client who, when told of the possible cancellation of the X-Trail programme at Nissan, described the current situation as "catastrophic for the industry", adding that he could see vehicle production falling, to 1m annually, a level last seen in the midst of the financial crisis a decade ago.

26. See: https://www.bloomberg.com/news/articles/2018-10-03/japan-waves-goodbye-to-u-k-as-gateway-to-europe-post-brexit

All references above were accessed/checked on January 29, 30 or February 3 and 8, 2019.

Academic reference cited:

Oliver Caudot and Jean Gourdon, *Non-tariff measures, preferential trade agreements, and prices: new evidence, Review of World Economics* (2016), 152, 277-249; available at https://econpapers.repec.org/article/sprweltar/v_3a152_3ay_3a20 16_3ai_3a2_3ad_3a10.1007_5fs10290-015-0242-9.htm

About the contributor

Ian Henry is managing director and owner of AutoAnalysis. He has more than 30 years' experience of consulting and analysis in the automotive industry, advising major car companies, suppliers, financial sector companies and government bodies in the UK and across Europe. He is a graduate of Cambridge University and has post-graduate degrees from Henley Management College and King's College, University of London. He specialises in production forecasting, supply-chain and competitor analysis. He is a regular contributor to leading automotive business media, notably www.just-auto.com.
www.automotivemanufacturingsolutions.com
and
www.automotiveworld.com.

Chapter 4

No light at the end of the production line

Brexit causes numerous and complex problems for the UK automotive sector, says Vicky Pryce, and there appears to be no easy answer to whatever happens

The UK automotive sector has been one of the great successes of the last decade. In 2016 it accounted for some 4 per cent of the economy, employed just under 900,000 workers, had an annual turnover of some £82bn and was one of the UK's main exporters with some 80 per cent of its output destined for abroad.[1]

It is one of the most productive sectors of the economy, investing some £3.6bn annually. Output had picked up again after the serious slowdown during the financial crisis with competitiveness bolstered by a weaker exchange rate and annual production rose to 1.7m cars in 2016, the highest number since the record year of 1972 when 1.9m cars left UK factories. There were some 2.7m engines built in the UK that year.

Yearly number of passenger cars produced in the UK *Source- Statista*

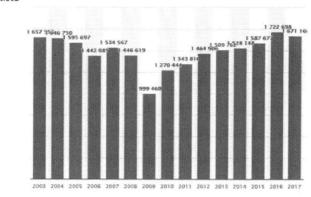

Output though fell back by 3 per cent in 2017 and is estimated to have fallen by over 9 per cent in 2018 as a whole to just 1.52m passenger vehicles. Latest monthly data, as I write, suggests a near 20 per cent fall year on year in November 2018 after a 10 per cent reduction in October, with UK car exports dropping despite the devalued Brexit pound. This has partly reflected a slowdown in the world economy and concern about diesel cars following the emissions scandal in Germany and no real clarity on how the UK government intends to proceed in this area.

But Brexit and the uncertainty it has created as to the future trading environment and the increased likelihood of no deal have been mentioned by the industry as major factors in the slowdown and the increased pessimism in the sector. Indeed The Society of Motor Manufacturers and Traders (SMMT) referred to Brexit as 'the most significant threat to the competitiveness of the UK automotive sector in a generation'.[2]

The consumer, now over-indebted again after a massive monetary relaxation by the Bank of England to assist the economy after the Referendum, has been a major part in this slowdown. Business optimism and consumer optimism have been in decline in recent months and private investment across the economy as a whole fell again in the three months to September (2018), the third quarter of consecutive decline.

Business optimism low

Manufacturing in the UK, which was in recession in the first half of 2018, enjoyed a brief recovery in the summer but export demand appears to have weakened since and there have been renewed monthly falls since. The car sector has been particularly badly hit. New cars registered in the UK have now fallen for two years – by some 6 per cent in 2017 and by a further estimated 7 per cent in 2018.[3]

Investment in the sector has been on a declining trend in the last few years. Latest data suggest that it halved in 2018 alone with the result that car manufacturers invested just £589m in the UK compared to £2.5bn in 2015.[4] And business optimism in the sector is at its lowest it has been for 27 years as Brexit uncertainty is taking its toll.

That uncertainty has increased since the UK Parliament voted down Theresa May's Withdrawal Deal with the EU in early January 2019 and in favour of reopening the negotiations on the backstop – which could lead either to a no deal, with which industry would find difficult to cope, or an extension to Article 50, which would prolong the conclusion of any firm arrangement for exiting the EU. Jaguar Land Rover(JLR), which was in the red last year, has already announced plans to cut 4.500 jobs, many in the UK amongst worries that a no deal Brexit will cost it an extra £1.2bn in lost profits per annum.[5]

In parallel the British Chambers of Commerce (BCC) has been reporting the development of serious skills shortages in many sectors including the car industry where EU workers make up at least 10 per cent of the UK automotive manufacturing sector labour force.[6]

With many EU citizens deciding to leave since the referendum those still arriving from the EU to work are not sufficient to stop a serious drop in net EU migration, which was just 74,000 in the year to June 2018, the lowest since 2014.[7]

And yet the sector remains a vital part of the economy and of UK manufacturing, accounting for 12 per cent of all exports.[8] With 54 per cent of UK car production going to the EU it is no surprise that developments on the Brexit front are worrying the UK car sector perhaps more than any other sector. Indeed, in his letter to Nicky Morgan, chair of the Treasury select committee on the implications of a no-deal Brexit, the Chancellor Philip Hammond singled out the car industry for mention among the sectors most likely to be affected by a no-deal Brexit.[9]

In a report of their Annual Executive Survey[10] in early January 2019, the EEF, the manufacturing association, and AIG Insurance found that Brexit was affecting manufacturing confidence negatively with the most feared issue being delays at the border.[11]

The end of frictionless trade EU would severely undermine the just-in-time manufacturing which has served the sector well. More than half of car components used in car assembly in the UK are produced elsewhere and a part can move multiple times between the UK and the rest of the EU to be improved/added to before it forms part of the final assembly over here.[12] Any end to this cooperation with continental suppliers – and their cooperation in turn with the 2,500 UK suppliers – would eat away at the UK sector's competitiveness, and with it the attraction of the UK to foreign investment.

And the regional implications would be severe. Work by Dennis Novy at the University of Warwick has calculated the unintended consequences for inflation of the Brexit vote as a result of sterling's depreciation since June 2016 and concluded that the North, Scotland, Wales and Northern Ireland, all areas with above average import content, suffered most in terms of the inflation increase that followed.

The study warns that even if sterling falls no further any tariffs imposed on EU imports after Brexit would exacerbate the inflationary impact felt so far while non-tariff barriers such as customs checks and red tape would ultimately also be damaging as they would feed into higher prices for British consumers. This would be particularly pronounced in areas where pan-European supply chains have been well established and end up being disrupted.[13] Looking at this analysis it is obvious that the car industry is likely to be one of the most affected by this – both in terms of costs and also consumer demand.

Multinational businesses already moving

Michael MacMahon from Oxford University[14] has looked at the impact of uncertainty over Brexit on business decisions. He concludes that "because the UK is so interconnected with other EU nations, uncertainty over Brexit has likely affected every aspect of firms' business environment". Worryingly, as he points out, a large number of companies appear to have only just realised the extent of the exposure of their business to membership of the EU – which allows them participation in global supply chains, access to labour across all skills levels, and access to all the services they need. Hence the recent rush to stockpile goods and components reported in the UK at the turn of the year.

In his piece for the 2019 Centre for Competitive Advantage in the Global Economy (CAGE) report MacMahon ends up lamenting the fact that firms are in fact facing what he terms 'policy-generated uncertainty' as government objectives often look completely undeliverable. One could of course add Brexit-generated policy paralysis which has had a much wider impact on the economy and which is likely to persist in some form whatever withdrawal deal may be voted through as it will leave many questions still unanswered.

Not surprisingly relocating activities seems to have started already. The SMMT's latest survey suggests that 12 per cent of firms are already relocating operations out of the UK and reducing headcount, as well as reviewing and actually altering logistics and other arrangements such as warehousing and stock adjustments, to be able to cope with Brexit.[15]

The reason to worry of course is that the shape of the sector has changed drastically over the past few decades. All the large manufacturers producing in the UK are foreign owned with only smaller specialists and luxury producers still mostly British owned.[16]

One of those is Aston Martin where the majority shareholding was bought back from Ford in 2007. The firm, having grown successfully, is still headquartered in Gaydon in Warwickshire[17] and has recently had an IPO and is now listed on the London Stock Exchange. Morgan, headquartered in Malvern, is another one but it only produces around 1,000 cars a year compared to, for example, the roughly current 135,000 annual production of Honda UK (Japanese owned, but closing) or the 200,000 per annum for Mini (owned by BMW, Germany).

The big foreign-owned companies, producing partly for the domestic market but also massively for export, can more easily adjust and move operations due to changes in the competitive environment. Their investment plans are being followed closely and Nissan UK (Japanese owned with an annual production of 500,000 units) was one of the first companies to be visited by Theresa May to reassure it on Brexit when she became Prime Minister after the Referendum vote. Despite the reassurances given so far the company has just announced that its investment plans for building the new X-Trail in the UK have been shelved[18], partly blaming the Brexit uncertainty for its decision.

Smaller component manufactures down the supply chain will be worried about these trends. And this concern about the supply chain as the BCC has identified will be intensified with the Honda announcement of its intention to close Swindon, which was already running at half capacity, at the end of 2021 and consolidate production in Japan.[19] It still plans to keep its European headquarters in Bracknell in the UK and the company did not attribute this decision directly to Brexit. Nevertheless uncertainty as to the final deal cannot have helped.[20] This means losses of 3,500 direct jobs and a possible further loss of an equal number again down the supply chain. Swindon workers are already blaming Brexit as a reason.[21]

It is worth noting that Honda had earlier announced plans to shut Swindon anyway for six days after the Brexit date of March 29 to ensure it could handle any issues with exports to the EU and also to address any concerns in relation to its imports of components post Brexit given that some 40 per cent of all components in Swindon come from the EU.[22]

In an autumn 2018 survey three quarters of SMMT members feared a no deal. The likelihood though is that the industry will suffer under any Brexit scenario. According to SMMT calculations a no-deal shift to the 10 per cent EU tariffs that would apply to UK cars as a third country, and 4.5 per cent tariffs on exports of components, would add some £1.8bn to the cost of exports and affect competitiveness adversely, while imports of cars from the EU would bear an extra cost of some £2.7bn, which would be hard for the sector to absorb at a time when consumers remain under pressure in their personal finances.

And one in ten had made contingency plans by relocating some of their activities.[23] As the SMMT points out, given that average import tariffs on car components under WTO rules are 4.5 per cent it makes sense for the UK to source its supplies from the UK or currently tariff free EU, keeping costs down.[24]

But though hard-line Brexit – a no deal – is feared most because of the rules of origin, customs controls, and delays, all Brexit scenarios would make the UK production more difficult and less competitive. And the industry has already faced Brexit-related costs due to uncertainty.

The same SMMT survey found that more than half of those who responded felt that their operations have already suffered due to lack of clarity over what future trading arrangement might end up being and nearly a third of respondents had already postponed or cancelled investment decisions in the UK as a result.

Even more worryingly, some 20 per cent reported having in fact lost business as a direct consequence of the Brexit uncertainty. And the EEF in its briefings has recommended that to ensure some continuity and minimise costs, if and when the UK leaves the Single Market, the manufacturing sector should be allowed to continue to participate in developing 'harmonised' product standards.[25]

Again, on behalf of manufacturers as a whole, it also pleaded for the current health and safety landscape to be maintained after leaving the EU so as to avoid creating extra costs and disrupting business.[26] There are also issues of regulatory uncertainty not just specific to the sector itself but also other cross sectoral legislation, which applies to all such as employment law and General Data Protection Regulation.

Cost-free Brexit is a dream

So what is in prospect? As Mike Hawkes, chief executive of SMMT has pointed out: "Frictionless trade as part of the EU Single Market and Customs Union has driven the success of the UK automotive industry so the fact we are leaving is already painful, and already causing damage."[27] Moreover any free-trade deal with other countries, such as the one mooted with the US, would necessitate a fall in import tariffs for cars coming to the UK from the US and that that could mean the loss of tens of thousands of car manufacturing and service jobs here.

A cost-free Brexit is a dream. Any added friction that augments costs to producers in what is a very tight-margin sector that has to meet tough emissions regulations and other safety standards will be affected by this.[28] Yes, some parts production may shift to the UK to avoid the friction created by Brexit, but the loss of the Single Market and possibly also being out of the Customs Union will make it more likely that any future expansion by the overwhelmingly foreign car makers currently operating in the UK will be in in continental Europe rather than here.

So the automotive sector is braced for quite a rough time ahead. A transition period would help but the concern over final trading arrangements will remain. What will emerge after the UK leaves is unclear. And then there is the question of financial support. Will the sums of money that are currently earmarked for transport, including the automotive sector, under the Horizon 2020 budget agreement and whatever succeeds it, still flow to UK car manufacturing?

High levels of R&D have been important in assisting competitiveness in the sector and it has been expected that the Horizon 2020 programme would continue temporarily, at least during a transition period, but if the transition does not happen then that would be in doubt.

Yet we know that in general the UK has tended to receive more back in science money than it has contributed through the EU budget process. If that source of future funding is cut or curtailed, will the Government, already having to borrow much more than it had envisaged before the Referendum, be able to step in? And will the Treasury have enough resources or the willingness to assist the regions that are likely to suffer most, both from the reduction in EU transfers and skilled labour and the damage that may be done to particular sectors such as the automotive industry by a badly executed Brexit? Don't hold your breath.

Notes

1. https://www.smmt.co.uk/industry-topics/uk-automotive/
2. https://www.bbc.co.uk/news/business-47055188
3. https://www.bbc.co.uk/news/business-46774053
4. https://www.bbc.co.uk/news/business-47055188
5. https://www.bbc.co.uk/news/business-46822706
6. https://www.smmt.co.uk/industry-topics/brexit/brexit-issue-papers/
7. https://www.ons.gov.uk/peoplepopulationandcommunity/populationandmigration/internation almigration/bulletins/migrationstatisticsquarterlyreport/november2018
8. Autocar12 August 2018
9. https://assets.publishing.service.gov.uk/government/uploads/system/uploads/attachment_data /file/735881/180823_CX_to_Chair_of_TSC_Nicky_Morgan_.pdf
10. https://www.insider.co.uk/news/brexit-manufacturers-eef-aig-survey-13821842
11. https://uk.reuters.com/article/uk-britain-eu-manufacturers/uk-factories-view-border-delays-as-a-major-brexit-risk-eef-idUKKCN1P1006
12. https://www.theguardian.com/business/2017/mar/03/brexit-uk-car-industry-mini-britain-eu
13. Did the Brexit vote lead to higher UK inflation? Dennis Novy, University of Warwick and Centre for Competitive Advantage in the Global Economy (CAGE) policy report, Social Market Foundation, 2019
14. The macroeconomics of uncertainty; Michael McMahon, University of Oxford and CAGE, Social Market Foundation, 2019
15. https://www.smmt.co.uk/2018/11/no-deal-brexit-catastrophic-blow-to-british-auto-industry-warn-businesses-in-new-survey/
16. https://www.quora.com/Are-there-any-remaining-British-owned-car-manufacturing-companies
17. https://www.telegraph.co.uk/news/uknews/1545308/Aston-Martin-back-under-British-ownership.html
18. https://www.theguardian.com/business/2019/feb/02/nissan-x-trail-uk-brexit
19. https://www.bbc.co.uk/news/business-47287386
20. https://www.theguardian.com/business/2019/feb/18/hondas-exit-is-based-on-many-factors-but-brexit-is-certainly-one
21. https://www.theguardian.com/business/2019/feb/18/workers-blame-brexit-for-demise-of-hondas-swindon-plant
22. https://news.sky.com/story/honda-to-stop-production-for-six-days-after-brexit-11603875
23. https://www.smmt.co.uk/2018/11/no-deal-brexit-catastrophic-blow-to-british-auto-industry-warn-businesses-in-new-survey/
24. https://www.smmt.co.uk/industry-topics/brexit/brexit-issue-papers/

25. https://www.eef.org.uk/campaigning/campaigns-and-issues/current-campaigns/manufacturing-and-europe

26. https://www.eef.org.uk/about-eef/media-news-and-insights/media-releases/2018/oct/nw-manufacturers-call-for-uk-to-stay-in-european-health-safety-regime-post-brexit-eefarco-survey

27. https://www.smmt.co.uk/2018/11/no-deal-brexit-catastrophic-blow-to-british-auto-industry-warn-businesses-in-new-survey/

28. https://www.smmt.co.uk/2018/11/no-deal-brexit-catastrophic-blow-to-british-auto-industry-warn-businesses-in-new-survey/

About the contributor

Professor Vicky Pryce is a Visiting Professor at Birmingham City University, a former Joint Head of the UK Government Economic Service (GES), and on the board of the Centre for Economics and Business Research (CEBR).

Chapter 5

Just-in-time listening required

Richard Burden MP and David Bailey explain why a hard Brexit combined with confused automotive policies would put the Midlands car industry at risk

Recent news that Jaguar Land Rover (JLR) will shed up to 5000 workers on top of the 1,400 lost in 2018 came as a huge blow for workers in the Midlands and showed the scale of the challenges faced by JLR and other motor manufacturers in the UK.

Nissan's u-turn on XTrail production coming to the UK was another illustration, linked in large part to declining diesel sales but with Brexit a complicating factor.

JLR itself faces a triple whammy of declining sales in China as the world's largest car market contracts after 20 years of breakneck growth, a massive shift away from diesels across Europe in the wake of the VW 'dieselgate' scandal, and Brexit uncertainty slowing the UK market.

That's even before the possibility of a no-deal Brexit throwing a spanner in the works of the whole UK automotive industry is added into the mix.

In the last week of January, MPs voted to tell the Prime Minister that, whatever else happens in the coming weeks, Britain must not be allowed to crash out of the EU without a deal – whether by the design of a so-called hard Brexit, or by default if the Article 50 process, which the UK triggered in 2017, expires on 29 March without any other arrangements being finalised in the meantime.

As we write this, Theresa May continues to resist ruling out a no-deal Brexit – as Parliament has called for – or to seek from the EU even a temporary extension of the Article 50 process to avoid a cliff edge on at the end of March.

Clear view from the manufacturers

Motor manufacturers have been clear about how damaging a no-deal Brexit would be for the sector. Describing as 'fantasy' claims that a no-deal Brexit can somehow be 'managed', Mike Hawes, Chief Executive of the Society of Motor Manufacturers and Traders (SMMT) warned of "immediate delivery shortages, disruption, additional costs and uncertainty."[1]

He is not kidding. The days of stocks of parts being stored in large warehouses are long gone and the industry relies on just-in-time deliveries to keep production lines going.

Honda's Swindon plant, for example, requires 350 trucks' worth of components to be delivered every day from Europe. Components arrive from EU suppliers within five to 24 hours. Honda and the other large automotive manufacturers with plants in the UK – like BMW, Ford, JLR, Nissan, Vauxhall and Toyota – know that even short delivery delays through the introduction of customs checks at the channel ports will bring production lines to a halt and cost them millions of pounds every day.

And that, of course is before adding in the cost of tariffs that would have to be imposed under the World Trade Organisation Rules that would apply following no deal. JLR has warned that trade tariffs and other costs would hit the company to the tune of £1.2bn a year.

Remember too that tariffs will not just be imposed on components imported to the UK. They will also raise the price of UK-made components that go to build cars assembled on the Continent, thereby hitting the competitiveness of the hundreds of small- and medium-sized companies here in the UK. Add to all this the impact of WTO tariffs adding 10 per cent to the cost of new vehicles exported from the UK to the EU and vice versa, and it is not surprising that automotive manufacturers and companies in their supply chains are putting investment decisions on hold.

Uncertainty leading major questions

In the medium term it could be even worse. Automotive companies make investment decisions about where and how they will produce new models years in advance, and Britain's Brexit uncertainty is already leading a number of firms to question whether they should move more of their operations overseas.

JLR is a company which has already invested billions of pounds in the Midlands economy and it retains an £80bn forward-investment programme that promises to put the Midlands at the centre of developing powertrains for its new generations of electric, connected and autonomous vehicles.

With a track record like this, there is little doubt of JLR's commitment to the UK. Even for a company like this, though, there is a limit to how long Brexit uncertainties can be withstood.

As JLR chief Executive Chief Executive Ralf Speth warned in an interview with the *Financial Times* in 2018: "At the end of the day we're in a cycle plan that means I have to make a decision…. I can't just wait, wait, wait, and wait." Even though "we built up this company over eight years," Speth noted that "It can go down the river so quickly."[2]

Speth is not alone in his fears. Figures recently released from the SMMT report that inward investment into the UK's automotive sector halved in 2018. As the clock ticks down to 29 March those figures are a stark reminder of what is already happening and how much more is a stake if the UK leaves the EU without a deal.

Government policy doesn't help

While Brexit uncertainties are a key factor in the sands shifting under the UK's automotive industry, they are not the only one.

The downturn in the Chinese market for internationally-produced vehicles is serious and – whatever President Trump's trade wars have done to aggravate that downturn – this is not a problem of the UK's making nor is it something within the UK Government's power to fix.

The same, however, cannot be said of the third challenge facing UK auto – the dramatic drop in the sales of diesel vehicles. In January 2019 the SMMT reported that after 21 consecutive months of decline, manufacturers were now selling 38 per cent fewer diesels than they did just a few short years ago.

Public concern about the impact of diesel emissions on air pollution has spiralled since the start of 'dieselgate' in 2015 – an act of self-harm which dramatically undermined trust in the integrity of automotive companies across Europe.

It is also true that historically, the level emissions of harmful emission such as nitrogen oxides (NOx) and particulate matter should have been addressed by automotive manufacturers and governments alike a lot earlier. With poor air quality now estimated to account for 28,000-36,000 preventable deaths in UK cities every year, the public attention that is now being given to air pollution is certainly justified.[3]

Diesels are getting better

Road transport contributes significantly to poor air quality and reducing the use of older diesel vehicles will be an important contribution to tackling the air quality challenge. Another part of the picture, however, is that design improvements have transformed the environmental performance of diesel engines produced to the latest Euro 6 standards. Add to that the fact that most diesel engines produce lower levels of greenhouse gasses than equivalent petrol engines, and it becomes even more important for public policy to distinguish between different generations of engines as well as different engine types.

Unfortunately this is an area in which UK government policy has been muddled and has sent out confused signals. Perversely, the introduction by the Government of punitive and differential rates of road tax on new diesel vehicles has hit sales of the cleanest diesels while providing no incentives to motorists to replace older, dirtier models.

Public statements by ministers have also sent out mixed messages. Some such as Business Secretary, Greg Clark, have been careful to draw distinctions between different generations of diesels. Others – like Environment Secretary Michael Gove – have adopted a less-informed and scattergun approach.

There is no doubt that the future of automotive does not lie with internal combustion engines – whether of the diesel or petrol. Effective management of the transition is, however, vital and decimating the market for new diesel engines has brought with it damaging if unintended consequences – even for the protection of the planet. For example, the dramatic drop in diesel sales has contributed to the first aggregate rise in the greenhouse gases produced by new cars in more than a decade.

The industrial impact of failing to manage the transition threatens to be severe too, with UK engine plants of manufacturers like BMW, Ford and JLR all currently heavily dependent on diesel production. Ministerial mixed messages over diesel undermines the capacity of manufacturers to manage that transition

A successful transition also requires more clarity from the Government in support of both the production and take up of the electric and other alternatively powered vehicles that will be the future of the sector. But the messages from ministers have been mixed here too with recent reductions in plug-in car grants standing in stark contrast to the incentives offered to motorists to buy zero-emission vehicles in counties like Norway.

On the positive side, after too slow a start, efforts are now being made by the Government to mandate the expansion of the UK's vehicle charging infrastructure, but much more needs to be done both to expand on-street charging and to make sure pubic charging points work properly whatever electric vehicle is plugged into them.

Here too, though, Brexit is unlikely to help. After the UK. leaves the EU, electric car sales in Britain will no longer count towards European emissions targets, threatening to stymie the UK's fledging electric car market before it gets properly charged up.

Make no mistake, the innovative capacity and diversity that has made the UK automotive sector the success story it has become over the past decade remain in place. However, the fact that a number of major manufacturers have yet to confirm plans to build in the UK the next generations of models (that the UK should be in pole position to produce) sends out serious warnings signals that would be foolish in the extreme to ignore.

By reducing Brexit uncertainty through ruling out no deal, ending mixed messages over modern diesel and showing much more dynamism in supporting the transition to a connected, autonomous and alternatively powered automotive future, ministers could show they are listening.

Notes

1.   https://www.smmt.co.uk/2019/01/smmt-statement-on-brexit-vote/

2.   https://www.birminghampost.co.uk/business/business-opinion/comment-stark-jlr-warning-hard-14870403

3.   https://www.kcl.ac.uk/lsm/schools/population-health-and-environmental-sciences/newsrecords/air-pollution-could-cause-36000-deaths-a-year-in-the-uk.aspx

About the contributors

Richard Burden is Labour MP for Birmingham Northfield and chair of the All-Party Parliamentary Motor Group.

David Bailey is Professor of Industrial Strategy at the Aston Business School.

Chapter 6

The dangers of sleepwalking...

David Shepherd warns of how heading for the cliff edge when not wide awake can be the cause of many problems

If your car is hanging off a cliff, your best chance of survival is by carefully exiting the vehicle. Although this situation can be frightening, evacuating safely is possible with a calm head and cautious planning. Braking the car, shifting the car's centre of balance, and evacuating slowly and in unison is a safe way to exit the car in case of emergency. With time and patience, you'll be able to get yourself and any passengers out of the car and to safety.

Perhaps a bit flippant but this advice from *WikiHow* seems apposite. The same piece goes on to say:

Stay calm. Hanging on the edge of a cliff can be terrifying but the calmer you are, the clearer your head will be while you escape. If you start to panic, take a deep breath and try to focus on your surroundings while you prepare to evacuate the car. Try not to make any rash movements. Unless your car is currently rolling off the cliff, take a few minutes to plan your escape.

Over the years, the automotive industry has weathered many storms but this time it feels different – and that's because it is different. This time the storm is unique to the UK and, somehow, it feels like we are on an uneven playing field. This isn't part of a cycle we've seen in the industry before and needs to be treated differently.

The travails of the early 2000s

Many of us will remember the dramatic and drastic closure of Rover in 2005; the closure of LDV in 2009; Vauxhall's Luton car assembly plant in 2003; the Transit finishing at Southampton in 2013, which in many ways seemed to mean more than car production stopping in Dagenham in 2000;

and others. As many businesses relied on Rover, its closure had a significant impact on the supply chain – especially in the Midlands.

However, in spite of all those closures, the UK is now assembling as many cars as it used to in the heyday of the 1970s. And there are many success stories too: Mini; Honda in Swindon, where the site was nominated as the global centre for Honda Civic; Nissan that not only builds cars but also carries out all-important design and development; Toyota; and, of course, the fantastic growth of Jaguar Land Rover (JLR) under Tata ownership. And all of that is before we talk about manufacturing more than two million engines per year.

However, in my opinion, the supply chain has not fared as well. With local content now estimated at around 45 per cent, the UK supply chain produces half of what it used to.

Over the years, I have seen the supply chain shrink, with downturns and to some extent recovery. As with other vehicle manufacture, we have seen major UK suppliers replaced by overseas suppliers. The supply base here is the same as you will find in Spain or Poland – more or less as in any country where vehicles are produced.

But I can't help but question how the different tiers in the supply chain will prepare for Brexit – deal or no deal – and how they will adapt to work in a new trade order.

Many businesses are increasing inventory levels to assure on-time delivery and avoid penalties for any service level failings. We see this not only in the automotive industry but in others such as the pharmaceutical industry. At least, in the short to medium term, we will need to continue to import materials and components and accommodate the costs of tariffs.

Contingency planning

The UK Government is also contingency planning for a disorderly Brexit, putting aside £100 million and signing contracts with French and Danish shipping companies to mitigate congestion.

Although we face uncertainty about the future of the UK automotive industry, especially in the supply chain, we can prepare for the worst and anything else will be a benefit.

As an industry, we need to build on our strengths, hold on to the strong positions we have established in the market and look into new opportunities – simply stock piling and warehousing will not be sustainable. Risk analysis in the automotive industry is now more important than ever.

However, there are still areas of opportunity in technology. If we are unable to lead on cost, maybe we should look at how to lead in other ways. For years, the UK industry has focused too heavily on human capital, rather than automation and mechanisation – perhaps now is the time to look more closely into the capability of new technologies such as Industry 4.0 and ways in which to use them.

As an industry, we could be looking at integrating new technology to bring down the cost of manufacturing and, in turn, sustaining profit and adding value to the customer. The way to sustainable profit is through a long-term view leading on technology rather than a short-term focus on the balance sheet. Bosch is a great example of this philosophy.

Using our craftsmanship heritage

Closer to home, there is the example of a Midlands pressings company which has adapted its business to incorporate fine blanking, supplying components to the medical industry the size of a pinhead.

We have a strong heritage of craftsmanship and the UK is seen as a leader in luxury and premium brands such as Bentley, Aston Martin and JLR. Eight of the 11 Formula 1 teams are based in the UK – demonstrating our engineering capability. We are going from strength to strength in automotive battery technology and we will soon reap the rewards of the Stephenson Challenge, as part of the UK Industrial Strategy, in further supporting the development of electric vehicle technology.

The engineering and development base of the Midlands is world class, which is why automotive companies from around the world have established and continue to set up technical and development centres here.

Now, more than ever, Benjamin Franklin's adage of 'If you fail to plan, you are planning to fail' should be at the forefront of people's minds.

The approach of the automotive industry needs to shift – adopting the principle of 80/20 where the focus is on a more lateral way of thinking (80 per cent) with a strategy (20 per cent) that supports this. I believe this would result in a more imaginative approach to business development to solve some of the challenges Brexit presents.

In terms of business planning and processes, employees will face new ways of working and changed cultures. There will probably be a need for more administrative controls such as customs documentation and visas that will add to indirect costs. This will all place different pressures on businesses and indeed the workforce.

Changing the direction of thinking

We need to be considering different markets and customers – thinking laterally. Spreading the risk by increasing exports has to be up there as a priority and there is much help out there to do so.

Trade missions can be a great way to reach new markets and customers but I'm surprised by how little take-up there is of these initiatives. For example, I recently attended an event that focused on helping supply-chain companies to establish a presence in Eastern Europe. Here there are areas of opportunity of sustainability and growth that would not only keep a foothold in Europe but, also, make it easier to reach automotive markets in territories such as, Belarus – with its growing industry – and Russia.

It may well be that businesses diversify into other sectors such as aerospace where the risks associated with Brexit are lower. Business may benefit from looking into acquiring tech firms or services – spreading business risk through coopetition. Some businesses might want to consider partnering, in some form or another; thinking about investor readiness would be time well spent.

If we make the assumption that the manufacturing cost base is going to increase with more expensive inputs and outputs, the supply chain will need to look elsewhere for added value.

The maths doesn't work but we can consider value, in simple terms, to be represented by function over cost. If cost increases, then function has to be improved to maintain the value status quo. Adding value for the customer is the holy grail and could help offset the increase in cost base to avoid sacrificing margin. Especially, as calls to reduce prices are likely to increase – as are the targets for reductions.

There is still much uncertainty but I believe the industry will come out of the planning process stronger and wiser.

About the contributor

David Shepherd is Senior Business Development Manager at the Drive Midlands Automotive Partnership, which is working with key leaders in the Midlands automotive industry to plan for the years ahead. It will be hosting workshops, networking events, sharing data and information.

For more information about the Drive Midlands Automotive Partnership, please email David.Shepherd@wmgrowth.com, visit https://www.drivemidlands.co.uk/ or follow it on Twitter at @DriveMidlands.

Chapter 7

Multinationals use Brexit to hide missteps

Britain's automotive business will thrive after the EU exit, says Neil Winton. The Referendum vote is being used to mask the sector's poor strategy and vision

Don't listen to the shrill naysayers who predict a Brexit no-deal disaster. The industry has form in predicting calamity if its advice is ignored, and there is more than a suspicion that bluster is designed to cover up corporate missteps.

Britain's automotive industry is going to survive and thrive after Brexit and remain a haven for making cars. There will be a free-trade deal sooner rather than later, possibly helped by invoking the World Trade Organisation's (WTO) Article 24, which allows nations declaring their intention of negotiating free trade to continue using current arrangements for two years.

News that Honda of Japan was closing its main European factory in Swindon reignited arguments about the negative impact of Brexit on the automotive industry, but the company refused to budge from its position that the decision was based purely on long-term factors. The impact of the EU-Japan Economic Partnership Agreement, in force since February 1 and under which the 10 per cent auto import tariff will be gradually phased out over 10 years, is a curve-ball which might prompt Nissan and Toyota to look again at their British operations.

Even if the UK 'crashes out' and has to deal with the EU on WTO terms (like the nation already does with the majority of exports), the industry will do well, according to reputable economists. The canard that ports will jam and motorways come to a halt the day after Brexit is just that, from an orchestrated industry response that is fearful of any change to its cosy arrangements.

And the industry has form. Remember the hysterical bullying when Prime Minister Tony Blair sought to get Britain to join the Euro before 1999? If we didn't join, the dark ages awaited. Only ill-informed fools could fail to see this, we were told by the automotive industry. Try telling that now to the huge numbers of unemployed youths in France, Portugal, Spain, and Greece.

But this time around automotive industry leaders who have charged into the controversial Brexit political debate are not only clouding the great prospects of life outside the EU. Some are cynically using the uncharted waters ahead to cover up their own inadequacies.

Grinding axes

Lead players like Jaguar Land Rover (JLR) and Ford have axes to grind as they noisily try to persuade Britons to overturn the referendum decision to leave the European Union, or frighten them it into accepting a hopeless deal which would be like not leaving at all.

JLR has been the most outspoken. Ralf Speth, CEO of the British-based and Indian-owned multinational, has floated some scary projections if a deal is not agreed. Key industries will be destroyed by a so-called hard Brexit, or no deal. Thousands of jobs will be lost.

"Everybody can do the maths. It is very open and very transparent. It is not only at Jaguar Land Rover, but it is also an element for the supplier industry and export industry in the UK," Speth told *The Guardian* newspaper. Speth said the prospect of a cliff-edge break with the EU was 'horrifying'. "A hard Brexit will cost Jaguar Land Rover more than £1.2bn a year – it's horrifying, wiping out our profit, destroying investment in the autonomous, zero-emissions (world) we share," Speth said.

Ford of Europe recently said a no-deal Brexit would be catastrophic, and regularly tried before the Referendum in 2016 to persuade the British not to be so stupid. In its most recent intervention, Ford's US HQ said it might cost the company up to $1bn if Britain leaves without a deal. No details were given.

JLR didn't spot diesel demise

Both companies are eager to blame the unknowable circumstances of Britain after Brexit for their own failings. JLR is in big trouble, not because of Brexit, but because it overestimated its sales in China, invested too much money in new product trying to compete with the likes of BMW and Mercedes, and relentlessly relied on diesel engines to power its big SUVs while disregarding the political clamour to ban them. A year ago more than 90 per cent of JLR vehicles had diesel engines.

Ford Europe too has been losing money because it failed to notice the SUV boom in time. And the red ink is going to accelerate, according to a report in 2018 by investment bank Morgan Stanley.

"We forecast Ford Europe to post an accumulated loss of $3.6bn from 2019 through 2021, with increased losses each year. By 2021, our forecast of Ford Europe's (profit) margin is negative 4.5 per cent which we estimate would make Ford the least profitable (manufacturer) in that market," Morgan Stanley said. There was no mention of Brexit.

Why would these huge multinational companies go out on a limb to try and stop the democratically endorsed Brexit? Global operators like JLR and Ford love organisations like the EU because it allows great operational freedom, keeps unwanted competition out and taxes to a minimum, and means they can cosy up to lawmaking bureaucrats in Brussels to keep it that way.

Grossly exaggerated

Economist Roger Bootle, chairman of Capital Economics, concedes that if he was an automotive industry leader he might be a little alarmed about the short-term prospects of no-deal and the theoretical threat to the supply chain.

"But the industry is talking as though the only integrated supply chain is in the European Union. They (supply chains) are in Asia and all over the world. They are grossly exaggerating their case, at a time when the industry is facing huge problems unrelated to Brexit. They made appalling decisions relating to diesel engines and the electric car, but they say we are in a terrible state all because of Brexit," Bootle said in an interview. "There is a huge amount of trade done around the world into the Single Market despite these pesky barriers," he added. He now expects a free trade deal to be negotiated, helped by the WTO's Article 24.

Ruth Lea, economic advisor to the Arbuthnot Banking Group, has no time for industry whining. "Instead of exaggerating the difficulties they would face trading with the EU under the WTO's tried and tested rules, they should be embracing the future. Fifty-five per cent of our export trade is basically conducted under such rules already. Rather than 'crashing out' and/or 'falling off a cliff', it would be more of a 'leap into the familiar'," Lea said in an email exchange.

Lea said countries trading under WTO rules had made big progress in recent years to streamline trading. "Under the landmark Trade Facilitation Agreement, developed countries with adequate resources are expected to install state-of-the-art border systems in order that trade should not be impeded. Most countries now permit traders to submit their customs documentation electronically in advance of the goods arriving at the border," she said.

Streamlined, computerised borders are the norm. Virtually all submissions of the EU's own Single Administrative Document, for declaring imports and exports, are now made online, for example. This means that most trade arriving from countries that are members of neither the Single Market nor the EU Customs Union suffer little or no hold up at the border when entering the EU. There is no reason for this to change after Brexit. Streamlined, computerised borders are the norm," Lea said.

In the interest of fairness and balance, the auto industry should be allowed to make its case. "The vote against the Brexit deal on the table brings us closer to the 'no-deal' cliff edge that would be catastrophic for the automotive industry. All sides in parliament must work together to find a way forward and put the necessary mechanisms in place to prevent this happening and explore alternatives that protect our future," SMMT chief executive Mike Hawes said in a statement. "No deal must be avoided at all costs. Business needs certainty so we now need politicians to do everything to prevent irreversible damage to this vital sector," he added.

Professor David Bailey, Professor of Industrial Strategy at Aston Business School, agrees. "A hard Brexit would be nothing short of catastrophic for the UK car industry as the SMMT itself recognises. Output will fall, jobs will be lost and – in my opinion – plants will close. UK auto needs a Brexit deal that secures free and frictionless trade with Europe, unrestricted access to the single market and the ability to hire skilled employees from overseas," Bailey said.

Riding out the storm

Peter Wells, Professor of Business and Sustainability at Cardiff Business School, agrees that uncertainty over Britain's post-Brexit arrangements is bad for its automotive business. The British car market is sliding and is preparing for hard times, but he is more optimistic for the future.

"However, the longer-term picture is by no means so bleak. In the automotive industry, and especially in vehicle assembly plants, the investments are very durable. It is expensive to close an assembly plant, in fact almost as expensive as to build a new one. As a result, it is likely that vehicle manufacturers will have to ride out the storm until a clear view of the trading arrangements can be determined," Wells said.

"Moreover, over half of UK new car exports are to markets outside the EU, a share that has been steadily increasing over recent years," he said. Wells points out that the UK auto industry has also benefitted from the steady decline in the value of sterling on foreign exchange markets, allowing it to be more competitive.

Capital Economics' Bootle hopes that the British government will take the opportunity to reform its economy by slashing regulation and cutting taxes and costs. "I'm hoping that if we have a full Brexit the government will take the opportunity to modernise our economy, while it may be that at the same time members of the European Union will be moving in the other direction. This would mean Britain will still be a haven for automotive manufacturing," Bootle said.

About the contributor

Neil Winton is a Forbes contributor, writing about the global automotive industry. He worked for Reuters for more than 30 years with assignments including Brussels, New York and Toronto, and Science and Technology, European Automotive, and Millennium Bug 2000 beats. He also wrote the European Perspective column for the *Detroit News*, and publishes the automotive analysis website www.WintonsWorld.com. Contact him at neil.winton@btinternet.com or twitter @wintonsauto

Chapter 8

Policy will matter

David Bailey thinks ahead to what short- and longer-term strategies will be needed post Brexit to support UK auto

The state of the UK's automotive industry (hereafter 'UK auto') seems to capture the zeitgeist of current concerns over Brexit in general and no deal in particular.

Partly the focus on auto arises because, until recently, it had been seen as something of a star performer in the UK economy, unlike many other manufacturing sectors. Output increased by 60 per cent over 2010-2016, with in excess of £8bn worth of investment in the industry over the same period.[1]

The sector supports some 850,000 jobs in total in the UK. This upturn benefitted regions, such as the West Midlands, which had struggled with deindustrialisation, plant closures and the legacy of the global financial crisis.[2]

There are a number of reasons for this automotive industry success up to 2016: the quality of the skills base, cooperative working between unions and management, strong links with universities, a supportive industrial policy, a shift upmarket leveraging UK innovation and design, and so on.

Instrumental in making a success of these was – critically – the UK being part of the EU Single Market, with significant foreign direct investment (FDI) by multinational automotive original equipment manufacturers (OEMs) coming to the UK to access and serve the Single Market.

What's more, the auto industry in the UK is seen as having benefitted from the UK's EU membership through the EU cutting trade deals with the rest of the world, in the UK influencing EU regulations affecting UK auto, and in accessing skilled workers and European research funding and networks.[3]

Fasten your seatbelts

And there's the rub. Such factors will now be re-evaluated by automotive firms in the wake of Brexit, and here a bumpy ride awaits.

Of late sales and output in the sector have fallen (down 7 per cent and 9 per cent respectively in 2018), linked in part to a cooling of the Chinese economy, a shift away from diesels, and uncertainty over Brexit affecting confidence. Investment has declined markedly (down by 80 per cent since 2016) as OEMs and suppliers postpone investment decisions, awaiting clarification on the form of Brexit and the UK's future trading relationship with the EU.

With some major OEMs' model replacement decisions looming (e.g. the Vauxhall Astra model at Ellesmere Port), this raises the risk of assembly shifting elsewhere if uncertainty continues. And even if Parliament does support a Brexit deal, the transition period offers only short-term clarification on trading on existing terms; longer term, a new trading arrangement needs to be agreed, again raising issues of uncertainty for the sector.

The industry is also seen as heavily exposed to the form of Brexit given the nature of its integration into automotive value chains across Europe. Not only do fine-grained supply chains cross borders many times, so possibly being affected by tariffs in the event of no deal, but the nature of modern just-in-time (JIT) delivery means that assemblers keep limited stocks to keep costs down. Components flow across the Channel on a daily basis as they are needed and in the order they are required.[4] Even minor disruptions to customs arrangements could have a major impact on supplies and hence existing business models for auto assemblers operating in the UK.

A no-deal scenario is seen as an 'existential threat' to UK auto by the Society of Motor Manufacturers and Traders (SMMT), given exposure to tariffs and customs delays; Henry in this volume points to a production loss of at least 175,000 cars a year in the event of no deal. This is just the short-term hit, however; longer term we would likely see plant closures.

Longer term this immediate output loss would rise substantially if UK plants are shut in the wake of a no-deal Brexit. Ford Bridgend, Vauxhall at Ellesmere Port and Jaguar Land Rover at Castle Bromwich in the Midlands are three sites which are the most vulnerable in such a scenario. And note that there would be no coming back; capacity would be permanently reduced. There would be a scarring effect on output and employment. Don't expect a wave of Japanese, or Korean or Chinese investment to turn things around as in the 1990s. Japanese investors came to the UK to access the EU's Single Market, after all.

Deploy airbags

The shock to the auto industry of no deal would have profound impacts on UK auto, including its suppliers, workers and the places hosting such activity. Policy responses would need to draw on earlier experience in dealing with automotive shocks to cushion the blow – such as the Rover Task Force and the Automotive Response Programme in the Midlands in the wake of the global financial crisis.

Such previous experience has pointed to the need for a range of measures to anticipate and respond to shocks. Given that the production hit would cascade down the supply chain, business support would need to include help for otherwise viable firms through measures such as loan funds, temporary wage subsidies, diversification advice, and tax and rate relief.[5] Workers would need support in terms of training and retraining.[6] Places hit would need measures to remediate sites, improve connectivity and regenerate places, in turn raising questions over the degree of devolved powers to achieve this.

It is not at all clear that government is prepared for such wide-ranging policy interventions.

Take another road?

Of course, avoiding no deal should be a priority. Staying inside the Customs Union is essential for fine-grained automotive supply chains to run efficiently, and avoiding non-tariff barriers will be key longer term; the aim should be to effectively give UK auto something like access to the Single Market.

And beyond this, the UK will need to more than just a new trade relationship with the EU.

For example, Britain will need to do much more to create and develop its own skills given that one in 10 manufacturing workers in the UK come from another EU country; this means developing better systems for education, skills training, and re-training as part of a wider industrial policy, and one which is determined much more locally than it is now.

Sadly, on the latter – despite much early hype – the Theresa May Government's industrial strategy was something of a damp squib, seemingly killed off by a hostile Treasury that was anti-intervention and anti-devolution to the regions. So much for joining up industrial policy with 'place'.

But there is something to build on. What's been particularly encouraging over the last decade has been the work of the Automotive Council, which started at the end of the Labour administration and which developed under the Coalition Government (thanks to Vince Cable) into an effective body in fostering public-private cooperation.

The Council's work has, for example, set out clear priorities for key automotive technologies that need to be developed (such as on powertrains, light weighting and intelligent mobility) which has both aligned government support and funding and has underpinned business confidence and investment.

The Council's work was backed up by a range of (modest) interventions to boost skills, rebuild supply chains, and encourage investment in the industry, such as through the Regional Growth Fund, the Advanced Manufacturing Supply Chain Initiative, the Manufacturing Advisory Service (MAS), and MAS's Tooling up Fund to support investment in tools in the supply chain.

Sadly, most if not all of these policy interventions were scrapped by Sajid Javid during his time as Business Secretary. That was a big mistake as where policy was reasonably well developed, it really did make a difference. And while the subsequent industrial strategy under May and Greg Clark put in place 'sector deals', the funding on offer was a fraction of previous support.

Route guidance

So going forward, what is to be done?

Firstly, the work of the Automotive Council should be continued but backed up with far greater resources, to support innovation, skills development and supply chain building.

Secondly, sector and place need to be combined: the Government needs to look again at the degree of devolved powers. It will need to return to development bodies that can intervene more widely and strategically at a regional level, and do 'smart specialisation' through regional-level industrial policies. Combined authorities may be one way to do that (in cities at least). Beefing up the local growth hubs to fill the vacuum left by the abolition of MAS could be part of this 'combined authority plus' model, as would devolution of skills funding to the regional level.

Thirdly, there is much more that the Government could be doing in really trying to 'rebalance' the economy and reduce Brexit-induced uncertainty, for example by stimulating investment in manufacturing such as through enhanced capital allowances, by resurrecting something like the Advanced Manufacturing Supply Chain Initiative (preferably on a much wider scale), and by plugging funding gaps for small firms in the supply chain.

Fourthly, there is a need to support to modernise and reorientate the sector so as to find new development paths[7]; much more policy dynamism is needed to support the transition to a connected, autonomous and alternatively powered automotive future, as Burden and Bailey note in this volume.

Finally and more broadly, there is a strong case for UK industrial strategy to be afforded an institutional status similar to both UK monetary and fiscal policies. At the very least, it should be the subject of regular strategic long-term reviews. By giving it that sort of priority, the new government would send out the kind of powerful message that British industry and foreign investors need to hear given recent uncertainty.

None of this is a panacea for a messy Brexit. But regardless of the form of Brexit, a more interventionist industrial policy will be required for UK auto going forward, building on the public-private cooperation that has been developed so well over the last decade, and in contrast to what some deregulation-minded Brexiteers might suggest.

Notes

1.   SMMT (Society of Motor Traders and Manufacturers) (2016) 2016 Sustainability Report. London: SMMT.

2.   Bailey, David and Nigel Berkeley (2014) Regional Responses to Recession: A Case Study of the West Midlands, Regional Studies, 48(11), 1797-1812.

3.   KPMG (2016) Brexit: How would business vote? London: KPMG.

4.   Bailey, D, McCann, P. and Ortega-Argiles, R. (2018) Could Brexit spell the end for "just-in-time" production?, Prospect Magazine, 5th April 2018.

5. Bailey, David and Nigel Berkeley (2014) Regional Responses to Recession: A Case Study of the West Midlands, Regional Studies, 48(11), 1797-1812.

6. Bailey, David and Alex De Ruyter (2015) Plant Closures, Precariousness and Policy Responses: revisiting MG Rover ten years on, Policy Studies. 36(4), 363-383.

7. Bentley, Gill, David Bailey and Daniel Braithwaite (2017) Resilience, adaption and survival in industry sectors: remaking and remodelling of the automotive sector, in Williams, N and Vorley, T, ed.s, Creating Resilient Economies. Cheltenham: Edward Elgar.

About the contributor

David Bailey is Professor of Industrial Strategy at the Aston Business School.

Chapter 9

The fight is on

Automotive workers will not see the industry they rebuilt destroyed by Brexit say Steve Turner and Ben Norman of the Unite union

When Tory Brexiteers think of manufacturing, if they do at all, they likely picture a factory not dissimilar to South Marston in Swindon as it was in the 1940s. Several hundred workers applying canvas to the wood and rubber sourced from empire to make Spitfires.

The reality of South Marston today, as any Unite member there can tell you, is one built over 30 years of work and £2bn of investment. Honda Swindon is a global production hub for the Civic, an operation sustained by a daily delivery of 350 truckloads components suppled without border friction. The result is a plant capable of making 600 new cars every 24 hours, with 85 per cent exported worldwide from Bristol and Southampton. Swindon's Spitfires may not patrol the Channel, but one in five of America's bestselling cars are made there.[1]

The UK's automotive sector stands or falls with its workforce, and Honda is no different. The site employs more than 3,100 workers in manufacturing and logistics, and some15,000 more in its UK-based supply chain. A strongly unionised site, our members have faced significant challenges, but their achievements are self-evident. Through collective bargaining Unite shop stewards and members secure wages and conditions 41 per cent higher than the average for the South West, a region too often in the relegation zone of such league tables.[2]

The impact of a plant closure on a region is not without precedent. Two years after the closure of the Rover plant at Longbridge in the West Midlands, Unite's predecessor union Amicus surveyed 1,700 former workers to reveal that 23 per cent were still unemployed, while the 20 per cent who had found work were only earning the equivalent of the then minimum wage. Many more reported that there was simply no financial provision for retraining.

A complex crisis

The crisis now facing our members at Honda is complex, but no one doubts that Brexit and its uncertainty is the context. It is not by accident that Honda's arrival to Swindon in 1985 coincided with the prospect of falling tariffs and other barriers to Europe, just as its threatened closure looks to coincide with their return. More than 40 per cent of those Civic components are sourced in Europe.

At the time of writing the UK is on the brink of a politically driven hard Brexit and the two lost years of uncertainty have taken a hard industrial toll. The sector had anticipated surpassing the 1.92m production mark reached in 1972. Instead Brexit uncertainty takes its place alongside global overcapacity, the impact on sale of anaemic economic growth, and a government-caused crisis over diesel.

Alongside Honda and the closing Schaeffler plants of Llanelli and Plymouth, our stewards face cancelled investment at Nissan, redundancies at Jaguar Land Rover and Vauxhall/PSA and a looming threat of 'global realignment' from Ford. Seldom has an industrial success story turned sour so quickly or so needlessly. Unite will not see it go unanswered.

For Unite, this is a self-inflicted injury caused by a government whose ideologues utterly fail to understand modern manufacturing. We have not forgotten the words of Professor Patrick Minford who claimed the automotive industry would need to be 'run down' after Brexit.[3] Hastily deleted tweets do not hide the weight his words still carry in the Department of International Trade.[4]

Trade unions have not forgotten the price working people paid when Thatcher's Government last implemented them. We understand what it would mean for 856,000[5] automotive workers today.

Members across the sector

Of Unite's 1.3m members half a million work in manufacturing, while we are also the largest trade union in transportation and logistics. This means Unite members are found on every step of any given vehicle's production cycle; from the steel blast furnaces, to the OEM production lines, behind the wheel of the motorway-bound car transporters and on the quaysides shadowed by international containerships. It is with that vantage, guided by the collective insight of our members, that Unite has taken responsibility for being the leading voice within the trade union movement on Brexit.

Our starting position has been to acknowledge and respect the 2016 Referendum result and accept that the plebiscite placed dividing lines in our workplaces, no less than it did between towns and shires or between generations. Our aim has been to seek unity between those members who voted for Brexit and those who opposed it by standing firm on the central principle that however people voted no one voted to lose their job.

Through two years of debate and analysis Unite has developed a position which could not contrast more with the Government's ill-fated red lines. Unite is calling for the fullest possible access to the Single Market, which we believe could be negotiated on the basis of firm commitments from the UK on non-regression of workers' rights, standards, social, consumer and environmental commitments.

We also support the UK entering into a customs union with the European Union on the basis of securing barrier-free trade and having a stronger collective voice on future trade deals. We see a customs union not only as vital for our automotive, aerospace, metals and food industries, but as a guarantor of the economic integrity of the island of Ireland and the peace that sustains.

The Withdrawal Agreement and accompanying political declaration negotiated by Theresa May – in the broadest sense of that word – fell short of these demands and the other key industrial tests adopted by our 2018 Policy Conference following hours of democratic debate.[6]

Strategy required

Securing a Brexit which defends jobs has been an immediate and pressing need, but that alone will not solve the wider problems facing the automotive industry. That is why Unite has argued that Brexit in any form can only succeed if it is accompanied by an ambitious industrial strategy.

Such a strategy cannot be concocted and implemented from Whitehall alone, especially not by civil servants all consumed by Brexit. It must be the creation of government and industry, which genuinely consults with and meets the needs of automotive workers and is guided by the research and analysis of experts as showcased in these pages.

As our contribution, Unite has brought together shop stewards from all corners of our industry. These include the production plants of JLR, Toyota and Nissan; the engine plants of Ford, Toyota and BMW; the tier 1 suppliers of Cummins, Dana Axels and GKN Driveline, and heavy vehicle producers of Alexander Dennis, Wight Bus, Vauxhall and Leyland DAF.

Together these leading stewards and reps debate, publish and campaign for the ideas and strategies needed if our industry is to meet the seismic changes of the coming decades, ranging from a just transition for diesel, to new investment needed for electrification and the industrial and social impact of automation.

The talent, creativity and ingenuity of automotive workers in the UK is beyond question; nor is their determination. Hard work, flexibility and even sacrifice rebuilt the industry. This will not be allowed to be thrown away by fantasists and ideologues. Unite will continue to campaign for the future of the automotive industry.

Collectively we will confront challenges as our members' face them, just as we will fight any attempt to force workers to pay for a crisis not of their making. Our members overall priority, as it is for everyone in these pages, is to keep the automotive sector where it deserves to be – the jewel in the crown of British manufacturing.

Notes

1. https://www.just-auto.com/analysis/the-us-has-a-new-best-selling-car-and-1-in-5-come-from-england_id183043.aspx
2. https://www.plymouthherald.co.uk/news/plymouth-news/how-much-average-person-south-1155418
3. https://www.tuc.org.uk/news/minford-view-brexit-and-exports-lacks-credibility-says-tuc
4. https://www.mirror.co.uk/news/politics/government-hurriedly-delete-tweet-after-11518202
5. https://www.smmt.co.uk/reports/smmt-motor-industry-facts-2018/
6. **https://unitetheunion.org/who-we-are/policies/**

About the contributors

Steve Turner is Assistant General Secretary of Unite and Ben Norman is Unite's Automotive Researcher.

Chapter 10

Carmageddon down under

There are some tough lessons to be learnt from the experiences of what was the Australian automotive sector, says Alex de Ruyter

Does the demise of the Australian domestic car industry in a free-trade environment offer any lessons as Brexit negotiations (at the time of writing in February 2019) continue to mean uncertainty for UK auto?

As the current drama around Honda's announcement of shutting down its Swindon plant by 2021, with the anticipated loss of 3,500 jobs (and knock-on losses in the supply chain) demonstrates, the future of UK auto is now highly uncertain as a hard Brexit looms. This only adds to pressures facing the industry arising out of a shift from diesel to electric vehicles and a slowdown in demand from the all-important Chinese market.

I focus on the departure of Ford from the city of Geelong in the state of Victoria, given its primacy as a key employer for the city for much of the 20th century, and briefly consider what measures were put in place by government to try and foster economic diversification and regeneration.

At this point, dear reader, I must confess a personal interest as I am originally from Geelong and worked in the Ford Casting Plant (for a very brief period of time) knocking 'slag' off engine blocks, and my father also worked at the factory.

The start of a new sector

The Ford Motor Company of Australia was formed in 1925 and production commenced in Geelong in that year (with other sites following in Adelaide, Sydney, Brisbane and Fremantle), originally producing the famous Model T.[1] Domestic production was a necessity to supply the Australian market, as the Government had banned the import of luxury goods (including car bodies) in 1917 during the First World war in order to promote domestic industries – and provide new employment for its (horse-drawn) carriage-makers.

Here was a key example of the policy of protectionism that was typified by tariffs, which was to endure until the 1970s. Following the Second World war, expansion of the domestic industry in 1948 saw the emergence of the rival iconic Holden brand (an Australian subsidiary of General Motors, also known as GMH) that soon dominated the domestic market.

This prompted Ford to develop and produce an Australian version of the US-designed Falcon in 1960[2] (with successive models also designed in Australia from 1972 onwards), specifically adapted to cope with the (often) harsh Australian driving conditions.

The rivalry between these two US-owned subsidiaries (producing big cars with six-cylinder and eight-cylinder / V8 engines and petrol consumption to match) continued through the 1960s, 1970s and 1980s, with the Ford Falcon and the Holden Kingswood and then Commodore (an Australian version of the Vauxhall Omega) vying to be Australia's most popular car (epitomised in a popular 1970s' television sit-com *Kingswood Country* with the lead character's characteristic line to his son, "You're not taking the Kingswood!"[3]).

The Japanese make inroads

However the 1970s also saw more economical Japanese vehicles begin to make inroads into the Australian market (with Datsun, now known as Nissan, being an early pioneer) as import tariffs were lowered, with a particularly notable tariff cut of 25 per cent in 1973 by the then Whitlam Labor[4] government.

Fierce opposition from domestic producers (notably GMH, which stood down 5,000 workers in response) saw this cut reduced to 15 per cent but it was notable at the time that the industry, which at its peak employed 100,000 people, was seen as being characterised by:

"Too many producers with extensive operations in multiple states, resulting in product proliferation, scale inefficiencies, and components industries that were forced into exceptionally short production runs, together with excessive and costly parts inventories."[5]

This combined with the rising costs of fuel would erode the popularity of the large cars such as the Falcon that had underpinned the profitability of the US-owned subsidiaries in Australia.

Japanese companies, namely Toyota and Mitsubishi (who took over the Chrysler operations in Tonsley Park in Adelaide in 1981), of course, had started production in Australia. However, increased trade liberalisation in the 1980s and 1990s (under both Labor and Conservative governments) emphasised measures to promote industry 'competitiveness' (the 1984 Button Plan of phased tariff reductions epitomised this for steel and textiles as well as automotive[6]).

This, combined with the emergence of the more cost-competitive 'Asian Tiger' economies, would combine to undermine fatally the logic of domestic production in Australia as the new century dawned. Thus, car production in Australia peaked in the 1970s (475,000 in 1970, which ranked Australia tenth in the world at the time) and declined more or less from there on.[7]

Large-scale production ends

By May 2013, Ford had declared its intention to close down its remaining production plants at Geelong and Broadmeadows (in Melbourne), following a cumulative period of successive losses. This had been preceded by Mitsubishi's closure of its Adelaide plant in 2008.[8] And GMH (December 2013) and Toyota (February 2014) followed suit, By the end of 2017 a period of some 70 years of large-scale vehicle production in Australia had come to an end.

For Geelong, the ceasing of manufacturing was particularly painful as, prior to the closure of Ford, manufacturing had provided (in 2012) some 44 per cent of the city's output.[9] The company, not only directly in terms of its own workforce but also in terms of dependent SMEs in the supply chain, was the most significant employer of the city's workforce. Whilst the Federal Labor Government in 2013 had provided a package of assistance measures to facilitate adjustment totalling some A$66m for displaced workers, these measures were principally focused on retraining.[10] Subsequent efforts have seen some manufacturing return to the site of Ford's former Geelong operations plant in the form of wind turbine assembly by Danish firm Vestas:

"The VREH will involve investment of approximately A$3.5m and directly employ over 20 employees. The project will train hundreds of local staff in wind turbine maintenance and see wind turbine component assembly in Australia for the first time in over 10 years."[11]

However, as evident, the number of staff directly employed in this is trivial. It remains to be seen whether this can fully compensate for the loss of vehicle manufacturing to the Geelong region.

The Australian experience hence points to the primacy of the state in fostering industrial policy and industrialisation. In this sense, the availability of abundant, cheap raw materials (a 'location-specific advantage' as economist John Dunning would have called it in his 'eclectic paradigm' of explaining foreign direct investment) combined with a protectionist state regime that offered incentives such as land packages meant domestic production was the only feasible way to supply the Australian market.

Proximity to a large domestic market is also an attraction for car producers. Australia did not have a large or integrated domestic market, being subject to a historical legacy of varying jurisdictions across the different states.

Furthermore, its trade liberalisation and integration policies with the Asia-Pacific meant that it made more economic sense for companies such as Ford and Toyota to now produce in the geographically proximate countries of Indonesia and Thailand and export complete vehicles to Australia. As such, the Australian experience also points to the transformative role of the state (at various spatial levels) as a driver of the structural changes all too readily solely attributed to globalisation.[12]

The parallels for Britain

For the Brexiting Britain, the parallels are striking. Like Australia in its protectionist days, domestic production is given a boost when the presence of external trade barriers acts as an inhibitor to exporting to the UK from a country of origin outside the EU; and the location-specific advantages of the UK with its flexible market environment *inside* the EU to date has made it an attractive location to be a production platform integrated with the rest of the EU.

Also striking is how the car industry in the UK, much like Ford was in Geelong, is particularly important for middling-sized towns/cities such as Coventry, Wolverhampton, Derby and Swindon (all in areas with a majority Brexit vote).

Stripped of these advantages post-Brexit and facing likely new tariff and regulatory barriers, the clear incentive – as Honda and Nissan are currently demonstrating – for multinationals will be to divest themselves over the coming period of production in the UK and reinvest elsewhere.

While state direction to foster the growth of high-technology sectors to compensate for the loss of vehicle production may soften the blow of plant closures and consequent job loss, the evidence to date suggests that the UK Government has yet to learn the lessons of rapid task-force response to such events (with associated place-based industrial policy Australia at least has state governments that are forced to consider regional development as part of their *raison d'etre*) and the need to re-skill workers to compete in emergent sectors.[13]

Epilogue

PS – I do not see anyone driving Falcons or Commodores anymore in Australia. Imported Japanese and European brands reign supreme...

Notes

1. https://www.abc.net.au/news/2016-10-07/timeline-ford-australia-ceases-production/7911742

2. https://www.abc.net.au/news/2016-10-07/timeline-ford-australia-ceases-production/7911742

3. http://www.adelaiderememberwhen.com.au/youre-not-touching-the-kingswood/

4. Labor, formally known as the Australian Labor Party, or ALP (American spelling convention used). The conservative parties in Australia have had various manifestations over the years, namely; Free Trade Party, Nationalist Party, United Australia Party and Liberal Party, generally as a coalition with the Country Party (now National Party).

5. http://theconversation.com/whitlam-made-the-case-for-reform-an-enduring-economic-legacy-33226

6. https://theconversation.com/whitlam-made-the-case-for-reform-an-enduring-economic-legacy-33226

7. https://en.wikipedia.org/wiki/Automotive_industry_in_Australia

8. See: Andrew Beer (2018), 'The closure of the Australian car manufacturing industry: redundancy, policy and community impacts', Australian Geographer, DOI: 10.1080/00049182.2017.1402452

9. https://www.thegordon.edu.au/sitedocs/skilling-the-bay/geelong-regional-labour-market-snapshot.aspx

10. https://www.abc.net.au/news/2016-10-07/timeline-ford-australia-ceases-production/7911742

11. https://www.rdv.vic.gov.au/news/wind-turbines-bring-manufacturing-jobs-to-geelong

12. Ibid. Beer argues convincingly that this process has been typified by the Federal Government in Australia prioritising overall growth and competitiveness over regional well-being, with state

governments and local authorities being left to deal with the "negative consequences of economic change".

13.   See Bailey in this volume.

About the contributor

Professor Alex De Ruyter is Director at the Centre for Brexit Studies at Birmingham City University and is a co-editor of this book.

Afterword

Driving over the cliff with the ideological Brexiteers?

John Mair

The car industry has been the bell-weather of the British economy for the last 50 years. When Ford sneezes, Britain catches a cold. Auto has been a huge success story in the Midlands and further afield. Car makers from Europe and Asia have flocked to the UK to take advantage of the Common Market.

Could Brexit – hard line or even a crash out of the EU – prove that 'Project Fear' from 2016 has proved to be 'Prophecy Fair'? The essays in this book have provided some answers, many of them negative.

The bubble of industry success has been based on big exports, ample supplies, regular model changes, the rise of company cars and ease of finance though PCP – the big-ticket item of buying a car has become a series of much more affordable monthly little tickets. Cars were almost driving themselves out of showrooms for many years.

There have been recent demand dampeners such as the collapse in the sale of diesel cars, thanks to their noxious effects being realised, and consumer distrust due to a series of scandals involving fiddled tests by manufacturers. But those jolts will be naught compared to the huge shock about to be delivered by Brexit.

On a personal level, I am an irregular visitor to the Oxford Motor Park which, like so many others, has grown up mushroom-like on the edge of the city. It has showrooms for most of the main marques. It used to be buzzing. Today, if I go into the Toyota showroom (my brand of choice), salesmen descend on me like locusts. Offers galore. Sales have not just gone down, they have tanked. Cut it how you like, Brexit and Brexit uncertainty have been bad news for the motor business. How long before an empty dealership appears on that showroom estate?

Two lone voices differ in this collection. Journalist Neil Winton argues that everything is going to be all right after Brexit, the auto multi-nationals just need to get their ideas into the right gear.

And Ross Clark, in Carmageddon, has a simple but telling analysis. It is not just Brexit but bad strategy that is the ruination of the British car industry – failure to spot the diesel disaster and failure to spot the rise of electric and hybrid vehicles.

In an analogy from an earlier era of technology, manufacturers were still making Betamax video recorders when consumers wanted DVDs and hard-drive memories. The early birds – like Tesla and the Japanese – have caught the electric worm. 'Catch up' is proving too little and too late.

However, their views do not carry weight with other contributors.

Richard Burden MP, chair of the All-Party Parliamentary Motor Group, argues that the historic home of the industry – the West Midlands – could become an industrial waste land post Brexit.

Economist Vicky Pryce, who used to advise British governments at the highest level, argues that this waste land will hit the niche makers such as Aston Martin and Morgan as well as the biggies. David Shepherd of DriveAutomotive looks at the car careering towards the cliff edge and tries to find ways of grabbing the steering wheel

Ian Henry of Auto Analysis does the sums and comes up with a neat figure. The costs of a hard Brexit will be £3.9bn to the car industry, interestingly one tenth of the £39bn 'Divorce Bill' proposed for the UK by the EU. Any gains from tariffs, on WTO or other terms, will not offset the costs. As I write, Porsche have warned future buyers to expect a 10 per cent price hike post Brexit. 'Project Reality'?

Staying in the realms of facts, figures and forecasting, LMC Automotive makes its living from just that – it forecasts that by the end of 2020, 400,000 fewer vehicles will be made in Britain, with the drop in output approaching 20 per cent. Toyota, for example, exports 90 per cent of its UK production to the EU. No more after 29 March 2019?

The doyen of auto economists, Professor David Bailey – one of the co-editors of this book – tries to craft a manifesto for the survival of the UK motoring industry. But even he cannot carve out a positive silk ear from the carnage. He believes plant closures and big lay-offs are inevitable.

In a very timely chapter Steve Turner and Ben Norman of Unite, which represents the Swindon Honda workers facing guaranteed redundancy in 2021, call for a change in strategy and for the UK to recognise the value of the EU Single Market. That cry may be too late...

Finally, from his homeland of Australia Alex De Ruyter – another co-editor of this book – paints a very dark picture of a possible scenario for the post-Brexit British auto industry. Successive governments there have failed to come to the aid of auto. Geelong and the Oz auto industry has simply tanked. Project Prophecy?

All in all, this book provides a depressing picture of the future. 'Project Fear' in 2016 may have been political hyperbole by David Cameron and George Osborne, but in 2019 and 2020 'Project Hope' for the UK motoring sector could prove to be just that.

False hope? That car and that cliff are coming closer together.

About the contributor

John Mair is one of the co-editors of this book. He is the series editor for the Bite-Sized Brexit books. John has edited 28 books on journalism and the media in the last decade.

Bite-Sized Public Affairs Books are designed to provide insights and stimulating ideas that affect us all in, for example, journalism, social policy, education, government and politics.

They are deliberately short, easy to read, and authoritative books written by people who are either on the front line or who are informed observers. They are designed to stimulate discussion, thought and innovation in all areas of public affairs. They are all firmly based on personal experience and direct involvement and engagement.

The most successful people all share an ability to focus on what really matters, keeping things simple and understandable. When we are faced with a new challenge most of us need quick guidance on what matters most, from people who have been there before and who can show us where to start. As Stephen Covey famously said, "The main thing is to keep the main thing, the main thing."

But what exactly is the main thing?

Bite-Sized books were conceived to help answer precisely that question crisply and quickly and, of course, be engaging to read, written by people who are experienced and successful in their field.

The brief? Distil the 'main things' into a book that can be read by an intelligent non-expert comfortably in around 60 minutes. Make sure the book enables the reader with specific tools, ideas and plenty of examples drawn from real life. Be a virtual mentor.

We have avoided jargon – or explained it where we have used it as a shorthand – and made few assumptions about the reader, except that they are literate and numerate, involved in understanding social policy, and that they can adapt and use what we suggest to suit their own, individual purposes. Most of all the books are focused on understanding and exploiting the changes that we witness every day but which come at us in what seems an incoherent stream.

They can be read straight through at one easy sitting and then referred to as necessary – a trusted repository of hard-won experience.

Bite-Sized Books Catalogue

Business Books

Marcus Lopes and Carlos Ponce
>Retail Wars
>>May the Mobile be with You

Maiqi Ma
>Win with China
>>Acclimatisation for Mutual Success Doing Business with China

Elena Mihajloska
>Bridging the Virtual Gap
>>Building Unity and Trust in Remote Teams

Rob Morley
>Agile in Business
>>A Guide for Company Leadership

Gillian Perry
>Managing the People Side of Change
>>Ten Short Steps to Success in IT Outsourcing

Saibal Sen
>Next Generation Service Management
>>An Analytics Driven Approach

Don Sharp
>Nothing Happens Until You Sell Something
>>A Personal View of Selling Techniques

Lifestyle Books

Anna Corthout
>Alive Again
>>My Journey to Recovery

Phil Davies
>Don't Worry Be Happy
>>A Personal Journey

Phil Davies
>Feel the Fear and Pack Anyway
>>Around the World in 284 Days

Stuart Haining
>My Other Car is an Aston
>>A Practical Guide to Ownership and Other Excuses to Quit Work and Start a Business

Stuart Haining
 After the Supercar
  You've Got the Dream Car – but is it Easy to Part With?
Bill Heine
 Cancer – Living Behind Enemy Lines Without a Map
Regina Kerschbaumer
 Yoga Coffee and a Glass of Wine
  A Yoga Journey
Gillian Perry
 Capturing the Celestial Lights
  A Practical Guide to Imagining the Northern Lights
Arthur Worrell
 A Grandfather's Story
  Arthur Worrell's War

Public Affairs Books

David Bailey, Alex De Ruyter, Neil Fowler and John Mair (Editors)
 Keeping the Wheels on the Road
  UK Auto Post Brexit
Eben Black
 Lies Lunch and Lobbying
  PR, Public Affairs and Political Engagement – A Guide
John Mair and Richard Keeble (Editors)
 Investigative Journalism Today:
  Speaking Truth to Power
John Mair, Richard Keeble and Farrukh Dhondy (Editors)
 V.S Naipaul:
  The legacy
John Mair and Neil Fowler (Editors)
 Do They Mean Us – Brexit Book 1
  The Foreign Correspondents' View of the British Brexit
John Mair, Neil Fowler and Alex De Ruyter (editors)
 The Case for Brexit
Christian Wolmar
 Wolmar for London
  Creating a Grassroots Campaign in a Digital Age

Fiction

Paul Davies
  The Ways We Live Now
    Civil Service Corruption, Wilful Blindness, Commercial
    Fraud, and Personal Greed – a Novel of Our Times
Paul Davies
  Coming To
    A Novel of Self-Realisation

Children's Books

Chris Reeve – illustrations by Mike Tingle
  The Dictionary Boy
    A Salutary Tale
Fredrik Payedar
  The Spirit of Chaos
    It Begins

Printed in Great Britain
by Amazon